As our nearest star, the Sun offers a unique opportunity to study
stellar physics in action. Following the success of his previous books,
Galaxies and *The Stars*, Roger Tayler presents the first full picture of
how studies of the Sun and the solar system help us understand stars
in general and other planetary systems.

Using mathematics appropriate for advanced undergraduate
students in physics, this textbook provides a broad and wide-ranging
introduction to the Sun as a star. Succinct derivations of key
results – such as the properties of spectral lines, the theory of stellar
oscillations, plasma physics, magnetohydrodynamics and dynamo
theory – are provided in a number of handy appendices, ensuring that
the book is completely self-contained.

Altogether, this is an invaluable textbook for students studying the
Sun, stars, the solar–terrestrial environment and the formation of
planetary systems.

The Sun as a Star

The Sun as a Star

Roger J. Tayler

University of Sussex

CAMBRIDGE
UNIVERSITY PRESS

Published by the Press Syndicate of the University of Cambridge
The Pitt Building, Trumpington Street, Cambridge CB2 1RP
40 West 20th Street, New York, NY 10011–4211, USA
10 Stamford Road, Oakleigh, Melbourne 3166, Australia

© Cambridge University Press 1997

First published 1997

Printed in Great Britain at the University Press, Cambridge

A catalogue record for this book is available from the British Library

Library of Congress cataloguing in publication data

Tayler, R. J. (Roger John)
The Sun as a star/Roger J. Tayler.
    p.   cm.
Includes bibliographical references and index.
ISBN 0 521 46464 1. – ISBN 0 521 46837 X (pbk.)
1. Sun.   2. Stars.   3. Astrophysics.   I. Title.
QB521.T39   1997
523.7–dc20   96-2862 CIP

ISBN 0 521 46464 1   hardback
ISBN 0 521 46837 X   paperback

Contents

Preface

Most stars are to a good approximation spherical and isolated and most of a companion book *The Stars: Their Structure and Evolution* was concerned with stars of that type. If we were able to observe the Sun from a large distance, we would regard it as an isolated spherical star whose properties were unchanging with time, unless our technology was so advanced that we could measure the small perturbation in its motion produced by Jupiter.

Because the Sun is so near to us, we can study the complex phenomena occurring in its surface layers, such as sunspots, solar flares and solar oscillations. We can use observations of oscillations and of solar neutrinos to obtain a better understanding of the solar interior and we can study the interaction of the solar wind with the magnetic fields of the Earth and other planets. A study of the properties of the planets and of the less massive members of the solar system, may provide clues concerning the formation of both the Sun and the planetary system.

In recent years it has become possible to study surface activity in stars other than the Sun, although in most cases the activity has to be much more intense for such study to be possible. Observations of star-forming regions and of young stars are also giving clues to the fine details of star formation and to the existence and frequency of other planetary systems. The purpose of this book is to ask how the study of the Sun and of the solar system may help us understand stars in general and how the properties of other active stars fit into an evolutionary picture.

The writing of this book was stimulated by two things. In 1988 I was invited to deliver the George Darwin lecture of the Royal Astronomical Society and I chose the topic *The Sun as a Star* because it was a subject which I wished to understand better. In 1992 a colleague Peter Christiansen and I introduced a new course for second year under-graduates at Sussex on *Space Physics and the Solar System*, with Peter teaching the space physics part of the course. Unfortunately Peter died suddenly following a heart attack in the Summer of 1992 and I became responsible for the whole course. With the permission of Peter's widow Sonja, I dedicate this book to his memory. Although this is not a conventional textbook, it should be suitable for courses at that level.

The problems discussed in this book involve much complex physics and mathematics, such as plasma physics and magnetohydrodynamics. The detailed derivation of results required in the main text has mainly been placed in appendices with the aim of making things easier for the reader.

I have benefitted over the years from discussions about the Sun and solar-type stars with numerous colleagues including in particular Andrew Collier Cameron, Douglas Gough, Moira Jardine, Carole Jordan, Leon Mestel and Nigel Weiss. In a short elementary book like this it is neither possible nor appropriate to apportion credit for every item of research reported but it should be clear that very little of the work described is my own. Where I have used the diagrams of other researchers, this is acknowledged in the text. The subject is too wide for me to claim to have a full understanding of all of it, but I have tried to make clear the uncertainties which I have. It is a rapidly developing field and the book is certain to be out of date in some respects by the time that it appears.

I am once again very grateful to Mrs Pauline Hinton for typing the book.

R J Tayler

Symbols

b	distance of closest approach
$\mathbf{b}$	unit magnetic vector, perturbed magnetic induction
$\mathbf{B}, B$	magnetic induction, magnitude of
B_i	partition function of ith state of ionisation
B_ν	Planck function
B_φ	poloidal magnetic field
$B–V$	colour index
c_H	hydromagnetic speed
c_s	sound speed
Ca XII etc.	notation for ionisation state of element
d	distance
d	deuteron
$\mathbf{D}$	electric displacement
e^-, e^+	electron, positron
$\mathbf{E}, E, E_\parallel$	electric field, magnitude of, component parallel to $\mathbf{B}$
E_r	energy of excited state of atom
f	velocity distribution function
$\mathbf{F}_{mag}$	magnetic force
g	surface gravity, acceleration due to gravity
$\mathbf{H}$	magnetic field
H_p	pressure scale height
H, K	spectral lines of ionized calcium
i	angle of inclination
I_ν	intensity of radiation
$\mathbf{j}$	current density
j_ν	emission coefficient
$\mathbf{k}, k$	wave vector, magnitude of
l	mean free path
l, m, n	quantum numbers, atomic states, oscillations
$L, L_s, L_{rad}, L_{conv}, L_{cond}$	luminosity, total, radiative, convective, conductive
L	scale of variation of magnetic field
m	particle mass

M, M_s	mass, total
M_U, M_B, M_V	magnitudes
n, N, n_i, n_e	number density, equilibrium, ion, electron
n_r, n_i	population excited state, level of ionisation
O B A F G K M R N S	spectral types of stars
p	proton
$P, P_\parallel P_\perp$	pressure, parallel, perpendicular to magnetic field
P, P_{rot}	period, rotation
q	charge on particle
r, r_s, r_g, r_c, r_E	radius, total, of gyration, coronal, Earth
r	rate of detection of neutrinos
$\mathbf{r}$	position vector
r, θ, φ	spherical polar coordinates
t, t_D, t_{Th}	time, dynamical, thermal
$T, T_e, T_s, T_{c\odot}, T_c,$	temperature, effective, surface, central (Sun), coronal
$T_e, T_i, T_{kin}, T_\parallel, T_\perp$	temperature electron, ion, kinetic, parallel, perpendicular to magnetic field
u, v, w	cartesian components of velocity
U, B, V, R, I	stellar magnitudes
$\mathbf{u}$	perturbed velocity
$\mathbf{v}, \mathbf{v}_0, \mathbf{v}_D$	velocity, unperturbed, drift
$v, v_r, v_\parallel, v_\perp$	velocity magnitude, radial, parallel, perpendicular to magnetic field
v_{th}, v_{esc}	velocity thermal, escape
X, Y	mass fraction of hydrogen, helium
Z_i	charge of ion in units of electron charge
α	helium 4 nucleus (alpha particle), mixing length parameter, pitch angle
α, β	coefficients in dynamo-equation
γ	photon, ratio of specific heats
Γ	effective ratio of specific heats
ϵ	energy release
η	resistivity
κ	opacity, coefficient of heat conduction
κ_ν	absorption coefficient
λ	wavelength, decay rate
λ_D	Debye length
μ	magnetic moment, mean molecular weight
ν, ν_c, ν_0	frequency, collision, line
ν_e, ν_μ, ν_τ	neutrino electron, muon, tauon
$\boldsymbol{\xi}$	perturbation vector
ξ, τ, λ, ψ	dimensionless variables in solar wind equations
$\tilde{\omega}, \varphi, z$	cylindrical polar coordinates

ρ, ρ_E	density, charge density
σ	collision cross-section, electrical conductivity
$\tau_c, \tau_c, \tau_D, \tau_H, \tau_R$	time collision, convective, decay, hydromagnetic, resistive
Φ	potential gravitational, coulomb
Ψ	wave function
$\omega, \omega_B, \omega_p, \omega_{pe}$	frequency wave, gyration, plasma, electron plasma
$\omega_{nlm}, \omega_{nl}$	stellar oscillation frequencies
Ω	angular velocity
I, II, III, IV, V	stellar luminosity classes

Some symbols are used with more than one meaning in order to accord with standard usage but this should not lead to any confusion.

Numerical values

Fundamental physical constants

a	radiation density constant	$7.55 \times 10^{-16}\,\mathrm{Jm^{-3}\,K^{-4}}$
c	velocity of light	$3.00 \times 10^{-8}\,\mathrm{ms^{-1}}$
e	magnitude of charge on electron	$1.60 \times 10^{-19}\,\mathrm{C}$
G	gravitational constant	$6.67 \times 10^{-11}\,\mathrm{Nm^2\,kg^{-2}}$
h	Planck's constant	$6.62 \times 10^{-34}\,\mathrm{Js}$
$k\,(k_{\mathrm{B}})$	Boltzmann's constant	$1.38 \times 10^{-23}\,\mathrm{JK^{-1}}$
m_{e}	mass of electron	$9.11 \times 10^{-31}\,\mathrm{kg}$
m_{H}	mass of hydrogen atom	$1.67 \times 10^{-27}\,\mathrm{kg}$
m_{p}	mass of proton	$1.67 \times 10^{-27}\,\mathrm{kg}$
ϵ_0	permittivity of free space	$10^{-9}/36\pi\ \mathrm{farad\,m^{-1}}$
μ_0	permeability of free space	$4 \times 10^{-7}\,\mathrm{henry\ m^{-1}}$
$\mathscr{R}$	gas constant (k/m_{H})	$8.26 \times 10^{3}\,\mathrm{JK^{-1}\,kg^{-1}}$

Non SI unit

eV	electron volt	$1.6 \times 10^{-19}\,\mathrm{J}$

Astronomical quantities

$L_\odot$	luminosity of Sun	$3.83 \times 10^{26}\,\mathrm{W}$
$M_\odot$	mass of Sun	$1.99 \times 10^{30}\,\mathrm{kg}$
$r_\odot$	radius of Sun	$6.96 \times 10^{8}\,\mathrm{m}$
$T_{\mathrm{e}\odot}$	effective temperature of Sun	$5780\,\mathrm{K}$
SNU	solar neutrino unit	$10^{-36}\ \mathrm{interactions\,s^{-1}\,target\,atom^{-1}}$
parsec	unit of distance	$3.09 \times 10^{16}\,\mathrm{m}$

1 Introduction

The Sun is responsible for our existence and it is therefore no wonder that many ancient civilizations worshipped the Sun as a god. There is no energy source, fossil or renewable, other than nuclear fusion and nuclear fission, which does not derive directly or indirectly from the Sun. Although attempts to use nuclear fusion as a power source on Earth are independent of the Sun, it is nuclear fusion reactions inside the Sun converting hydrogen into helium which provide the solar energy upon which we depend for our existence. Nuclear reactions of the heavy radioactive elements play only a minor rôle in the overall properties of the Earth, but studies of the abundances of these elements and their decay products in the Earth, Moon and meteorites enable us to estimate the age of the solar system and of the Sun, leading to the conclusion that the Sun has been luminous for about 4.6×10^9 yr. It was this age estimate which made it clear that the only known energy source capable of supplying the Sun's needs was nuclear fusion.

The Sun is our nearest star. It is so much nearer and looks so much different from other stars that the trick question 'what is the nearest star to us?' frequently gets the wrong answer. The distance from the Earth to the second nearest known star is more than 2×10^5 times the distance from the Earth to the Sun. Because of this very large ratio of distance, no star other than the Sun appears as anything but a point source of light. Stars are apparently faint and variations of properties over their surfaces cannot be observed through direct visual observations or photography. It is, however, natural to hope that, if we can obtain an understanding of solar properties, we shall be able to extrapolate our knowledge to those stars which we cannot study in anything like the same detail.

There have in the past been two approaches to the study of the Sun as a star and these approaches have tended to be followed by two distinct groups of astronomers, the students of stellar structure and evolution and the solar physicists. The first group regard the Sun as a source of radiated energy which appears to have been sensibly constant for periods of millions of years and not to have changed by an order of magnitude in the past 4.6×10^9 yr. The near constancy for the shorter times is easily understood once the origin of the radiated energy in nuclear reactions in the deep solar interior is appreciated. The Sun has such a large heat content that, if the central nuclear reactions were turned off, it would take about 10^7 yr before any knowledge of this reached the solar surface.

The Sun appears to be a hot sphere of gas held together by its own self-gravitation, which is kept hot by the steady nuclear reactions.

This view of a Sun with slowly varying properties has led to the study of spherically symmetric models of the Sun, which evolve slowly in time, and with corresponding models for the structure and evolution of all types of star. Although such investigations require much detailed knowledge of the physics of stellar interiors, they are conceptually simple. In this study of stellar evolution, the immeasurably greater knowledge which we have of the Sun compared with other stars is used to constrain theoretical ideas particularly inasmuch as they apply to stars which are not too different from the Sun in mass. The calculations predict that the Sun will be luminous until it is more than twice its present age but that before it dies it will become so luminous and large that the Earth will be engulfed. It is also predicted that the Sun will finally become the type of compact dead star known as a *white dwarf*. The idea that stars are basically simple objects to understand once led to an interchange of approximately the following form between two astronomers. 'A star is a simple object.' 'You would look simple at a distance of 10 parsecs*.'

Theoretical studies of stellar structure have been generally successful in providing an understanding of the observed average surface properties of the Sun. If the question is asked whether use of the established laws of physics, allowing for the uncertainties involved in a detailed application of them in calculations of stellar evolution, can lead to models of stars with the mass, chemical composition and age of the Sun, which also have the correct luminosity, radius and surface temperature, the answer must be a clear yes. There has however been one unsatisfactory feature in this comparison between theory and observation since the late 1960s. Theory predicts not only the surface properties of the Sun but also its interior properties. Included in these are the rates of the different nuclear reactions which occur in the conversion of hydrogen into helium near the solar centre and the consequent flux and energy spectrum of particles known as neutrinos which are emitted in some of these reactions. The conversion of four hydrogen nuclei into a helium nucleus releases two positrons and two neutrinos.

Neutrinos are very weakly interacting particles which escape essentially freely from the Sun but it is in principle possible for some of these neutrinos to be detected in a carefully planned terrestrial experiment. Ever since such an experiment was first mounted, the number of neutrinos detected has been fewer than that expected on the basis of a standard solar model which provides good agreement with the observed surface properties. This discrepancy could mean that we do not understand the internal structure of the Sun and this could have important implications for our understanding of other types of star. In fact, as I shall explain in Chapter 3, it is now generally believed that it is a lack of understanding of the properties of the neutrino which is causing the discrepancy. If correct, this would require a modification in the current standard model

---

* The parsec (3.09×10^{16} m) is the unit of distance most used by professional astronomers. It is the distance at which the mean radius of the Earth's orbit around the Sun subtends an angle of 1 second. It is 3.25 light years, where a light year is the distance travelled by light in 1 year. The nearest known star to the Sun is about 4 light years distant.

for the interactions of elementary particles, but would leave current ideas concerning stellar structure and evolution intact. This is an example of the very close relationship between developments in astronomy and in particle physics.

More recently it has been realized that there is another way in which the interior of the Sun can be probed. Like any other complex physical system, the Sun possesses normal modes of oscillation. It appears that many of these modes have been excited and can be observed. The properties of each mode depend on the internal structure of the Sun. Some are restricted to the surface layers but others probe the deep interior. If many different modes are observed, it is possible to put constraints on the variation of such quantities as temperature and density in the solar interior. This relatively new subject, helioseismology, promises to provide extremely useful information about the solar interior, which can be compared with predictions of standard models.

I now turn to the view of the Sun, which is the province of solar physics. This is one of extreme complexity with rapid spatial and temporal variability. The solar physicist studies the outer layers of the Sun in great detail. It is observed to have sunspots, which were first observed in the West by Galileo, and violent outbursts known as prominences and flares. It is also losing mass into space continuously, through what is known as the solar wind. The outermost layers of the Sun, the solar corona, are surprisingly at a very much higher temperature than the visible solar surface. This view of an *active Sun* is not in contradiction with that of the slowly varying Sun, which I mentioned earlier. The Sun does have steady average properties but upon these is superimposed a wealth of surface detail which requires a totally different explanation from the average properties. One particularly exciting discovery, which ties together the average Sun with the time variable Sun, is that of the solar oscillations and helioseismology, which I have just mentioned. Observations of rapidly changing surface properties can be used to obtain information about the slowly varying interior.

Because the Sun must be regarded as a typical star, it can be expected that many other stars, whose gross properties (mass, chemical composition, luminosity, surface temperature) are similar to those of the Sun, will also exhibit surface activity superimposed on their average behaviour. Because all other stars appear as a point source of radiation, it is almost impossible to study variations of properties across their surfaces. It would, as a result, be difficult if not impossible to detect activity at the solar level on most stars. More intense activity is easier to observe and it is now apparent that many solar-type stars do possess spots, flares, coronae and winds and the other attributes of the active Sun. In addition, there are stars whose gross properties are different from the Sun, which possess surface activity in more extreme forms.

As soon as there exist observations of a variety of stars with different degrees of surface activity, it becomes easier to ask what is the origin of this activity by seeing how its strength correlates with other properties of the stars. It is generally agreed that the existence of surface activity in the Sun, and in other stars with surface temperatures similar to or lower than that of the Sun, is closely related to the presence of a surface magnetic field and to the rotation of the star and to the presence of an outer convection zone. A major theme of this book is the relation of surface activity to the magnetic field, with the associated question of what it is that determines the strength of the magnetic field in the surface layers of a star.

The interaction of the Sun with the Earth is clearly of particular interest. At a time when there is great concern about our own interference with the environment, with a possible enhanced greenhouse effect leading to global warming and with worries about the hole in the ozone layer, it is also necessary to ask whether the Sun is likely to have any short term fluctuations in luminosity, which might influence the terrestrial climate. There is evidence from recent observations that the solar luminosity varies during the sunspot cycle, with the maximum luminosity occurring at sunspot maximum. Although the changes in luminosity in recent sunspot cycles have been very slight, there is also evidence that the *Maunder minimum* of sunspot activity in the 17th century was accompanied by a little ice age, at least in the Earth's northern hemisphere. This means that there is a possibility that erratic solar effects on the Earth's climate could be very important. Because even a change in the average terrestrial temperature of one or two degrees could significantly affect agriculture, a good understanding of what controls the sunspot cycle would be valuable.

The Sun interacts with the Earth through its particle emission as well as its electromagnetic radiation. The solar wind, which flows outward from the Sun through interplanetary space, encloses the Earth and its local magnetic field in a *magnetosphere*. At times of strong solar activity, the intensity of the solar wind increases and its interaction with the magnetosphere causes magnetic storms and aurorae. The magnetic storms disturb radio communication on Earth, which means that a better understanding of solar–terrestrial relations is of commercial as well as scientific interest. At all times the magnetosphere and the interplanetary medium close to the Earth provide a laboratory in which the subject of plasma physics, the study of ionised gases, can be studied in a manner which is not possible on Earth.

Another topic upon which it is hoped that studies of the Sun and the solar system might shed some light is that of star formation. Conversely, studies of stars which are currently forming might provide important clues to the origin of the solar system. Star formation is a field in which detailed observational information has only recently become available because of developments in infrared and millimetre-wave astronomy. One observational fact that has been known for a long time is that many if not most stars are members of binary or multiple systems. The Sun, in contrast, possesses a planetary system. Does this mean that there are two mechanisms of star formation, which produce binary stars and planetary systems respectively, or are there also stars which form individually? It has recently become apparent that there are many young stars which are surrounded by disks out of which planets may be condensing. Since this book went to press, reports have appeared of the discovery of Jupiter-like planets about normal stars like the Sun. Two planetary sized bodies have been detected orbiting a compact dead star, a *neutron star*. This is a puzzle because the neutron star is thought to have formed in a supernova explosion which might have destroyed planets. There is a possibility that they condensed out of the debris of the explosion. An important question is whether careful studies of the solar system can throw light on the problem of star formation and planet formation in general. There is obviously great interest in this because of the question of the existence of advanced life forms elsewhere in the Universe.

Chapter 2 is concerned with observations of the Sun, both of its average properties and of those more erratic properties known as solar activity. The discussion of the Sun is preceded by a brief account of the properties of other stars, as this will be needed when I discuss stellar activity. In Chapter 3 I discuss how theoretical ideas concerning stellar evolution influence our understanding of the overall properties of the Sun. In particular I give an account of the solar neutrino problem and of the way in which helioseismology can provide information about the invisible solar interior. Chapter 4 is the most demanding chapter because in it I summarise all of the ideas concerning the interactions between magnetic fields and fluid motions which are necessary for an understanding of the magnetic activity in the outer layers of the Sun and other stars and for a discussion of the interaction of the solar wind with the Earth's magnetosphere. Chapter 5 is concerned with the details of the magnetic activity of the Sun and Chapter 6 with activity of other stars. Solar-terrestrial relations are discussed in Chapter 7 and Chapter 8 is about the formation of the solar system, with comments on observations relating to the possible formation of other planetary systems. Chapter 9 contains some concluding remarks. The most difficult physical and mathematical concepts are discussed in Appendices.

It will rapidly become apparent that there is a vast amount of complex observational information relating to the Sun and also now to other active stars. It is quite impossible in a short book like the present one to describe the details of the observations and even less to attempt a theoretical interpretation of them. The most that I can hope to do is to provide a flavour of the observations and of the theoretical ideas which may lead to their understanding. Inevitably this involves telling some half truths but I have not deliberately told any untruths.

2 Observations of the Sun and other stars

Properties of stars

Before I discuss the Sun, I give a brief account of the properties of stars of other types in order to be able to place the Sun in context. I shall summarize these properties and not describe in any detail how they are obtained. Further details can be found in the companion book *The Stars: Their Structure and Evolution*. This book will frequently be referred to as *The Stars* in what follows.

The main properties which can be studied for any star, whose distance from the Earth can be measured or estimated accurately, are the total amount of energy emitted each second (absolute luminosity) and the spectral distribution of that radiation. Usually what is tabulated is the luminosity in a particular frequency range or equivalently the magnitude, which satisfies the relation

$$\text{Magnitude} = \text{Constant} - 2.5 \log_{10} (\text{Luminosity}). \tag{2.1}$$

The constant is in principle arbitrary but it is in practice chosen so that some normalising star has zero magnitude. The detailed spectral distribution of the radiation can also be approximated by measuring the luminosity in a number of frequency bands in different regions of the spectrum. The most commonly used system, the U, B, V system has three wide bands situated in the ultraviolet, blue and visual regions of the spectrum; since it was first introduced, the increased ability to do astronomy in the infrared has led to the introduction of further frequency bands in that region of the spectrum. The absolute magnitudes corresponding to U, B and V are called M_U, M_B, M_V. What is actually measured is an apparent magnitude, which is related by a formula similar to (2.1) to the apparent luminosity; this is the energy from the star crossing unit area at the Earth perpendicular to the direction to the star. This can only be converted to an absolute magnitude if the distance of the star is known. The constants in the two relations between luminosity and magnitude are related so that the absolute magnitude of a star is the apparent magnitude it would have at a distance of 10 parsecs. The apparent magnitudes are simply called U, B, V. The difference between two apparent magnitudes, such as $B-V$, is called a colour index. It is related to the colour of the radiation emitted by a star and also to its surface temperature. Note that $M_B - M_V = B - V$, unless absorption of starlight in the interstellar medium affects blue and visual radiation differently. Only if such absorption is unimportant is the conversion from apparent to absolute magnitude straightforward.

If stars radiated like black bodies, the spectrum of their radiation would be related directly to their surface temperature, T_s, through Planck's law

$$I_\nu = B_\nu \equiv (2h\nu^3/c^2)/(\exp(h\nu/kT_s) - 1). \tag{2.2}$$

Here I_ν is the intensity of radiation at frequency ν and h, k, and c are Planck's constant, Boltzmann's constant and the velocity of light. If this were the case, the colour indices would be uniquely related to the surface temperature. Stars do not radiate like black bodies. This is apparent in two ways in their observed spectra. The spectrum is interrupted by dark lines (*absorption lines*) at particular frequencies where the emission is greatly reduced. These spectral lines are produced by particular chemical elements which are absorbing prominently in the star's atmosphere and they provide the only information which we possess of stellar chemical composition. More rarely the spectrum is crossed by bright *emission lines* where the intensity is higher than average. In addition to the occurrence of spectral lines, the average emission from the star also deviates from the Planck curve. For some stars this deviation is not too great but for others it is very significant. In fact a star does not possess a precise well-defined surface temperature. It does not have a totally sharp edge and the radiation reaching an observer comes from a range of layers with different temperatures in the outer atmosphere of the star. Much of this book will be concerned with the variation in height of properties of the solar atmosphere. Theoretical astronomers introduce what is known as the *effective temperature* of a star. This is the temperature of a black body which has the same radius and luminosity as the star. Thus

$$L_s = \pi a c r_s^2 T_e^4, \tag{2.3}$$

where L_s is the luminosity, r_s the radius and T_e the effective temperature and a is the radiation density constant. There is then an approximate relationship between logarithm of effective temperature and colour index.

The Hertzsprung–Russell diagram

If M_V and $B-V$ are plotted against one another for nearby stars of known distance, a diagram which is schematically like fig. 1 is obtained. This is called a *Hertzsprung–Russell diagram* (HR diagram). Four main groups of stars are shown in the diagram, main sequence stars, giants, supergiants and white dwarfs. Most observed stars are main sequence stars with the second largest group being white dwarfs. It is explained in *The Stars* that the properties of stars vary during their life history. The Sun is currently a main sequence star but it is predicted that it will later be a red giant and finally a white

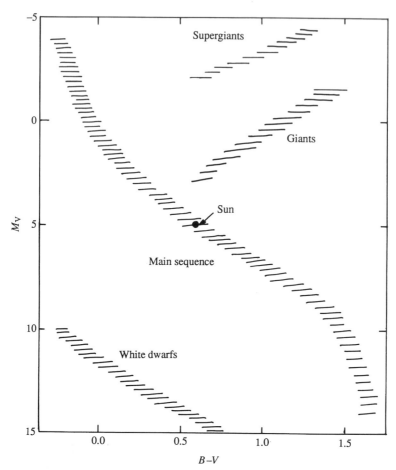

Fig. 1.   The Hertzsprung–Russell diagram for nearby stars. The visual magnitude M_v is plotted against colour index $B-V$ and most stars fall in four well-defined groups.

dwarf. Inside the main sequence band massive stars are at the top left and low mass stars at the bottom right. This indicates that main sequence stars obey a mass–luminosity relation, which is also schematically shown in fig. 2. The average slope of this relation between $\log L_s$ and $\log M_s$ gives

$$L_s \propto M_s^4. \tag{2.4}$$

Because of this high dependence of luminosity on mass, massive stars use up their supply of nuclear fuel, which provides their luminosity, much more rapidly than low mass stars. They therefore complete their life history much more quickly. The main sequence phase is the first major phase of stellar evolution. Main sequence stars much more massive than the Sun, which are observed today, must have been formed in very recent astronomical history. In contrast stars much less massive than the Sun would still be on the main sequence today, even if they were formed at the start of the Galaxy's history.

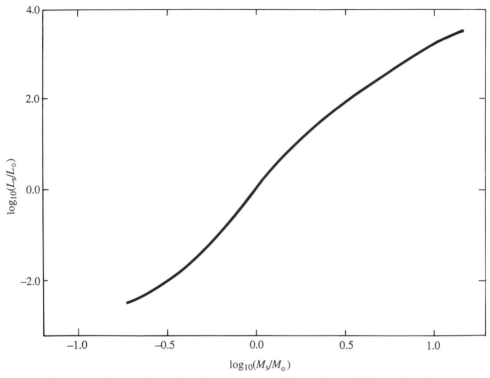

Fig. 2. The mass–luminosity relation for main sequence stars. The luminosity L_s is plotted against the mass M_s. $L_\odot$ and $M_\odot$ are the luminosity and mass of the Sun. Main sequence stars with accurately known luminosity and mass lie close to the curve shown.

The density of stars at different places in the HR diagram, which is not indicated in fig. 1, depends on two things, the distribution of masses of stars formed and the time spent in a particular evolutionary phase. Thus there are not many stars in the upper main sequence both because fewer massive than low mass stars are formed and because their main sequence lifetimes are shorter. There are also not many red giants because this evolutionary phase is much more short-lived than the main sequence phase. In fact luminous stars, such as those on the upper main sequence, supergiants and giants tend to be overrepresented in diagrams such as fig. 1 because they are more readily seen than low luminosity stars. There is one important group of stars which is not shown in fig. 1 with which I shall be particularly concerned in this book. These are known as pre-main-sequence stars. Before stars reach the main sequence they are not releasing energy from nuclear reactions in their interiors. As a result they can only release gravitational energy as they contract and heat up. I shall show in Chapter 3 that this occurs on a *gravitational timescale*, which is very short compared to the main sequence lifetime. In the case of the Sun, we believe that it will spend almost 1000 times as long on the main sequence as it took to reach it. If stars were being formed and evolving steadily, we should then expect there to be about 1000 main sequence stars for every pre-main-sequence star of about a solar mass, so that pre-main-sequence stars would be relatively rare. In any case there are not enough pre-main-sequence stars close to the Sun to be plotted in fig. 1. In this book I shall only be concerned with pre-main-sequence and main sequence stars and giants.

Spectral types and luminosity classes

Towards the end of the nineteenth century it was found useful to classify stars in terms of properties of their observed spectra into *spectral types*. The types were based on which elements were prominent in stellar spectra and it was originally thought that this represented the actual chemical composition of the stars. In the 1920s, following the work of M N Saha on the properties of hot gases in thermodynamic equilibrium, it was realised that this was incorrect. Spectral lines of a particular element can only be produced in a stellar atmosphere if the element is in the right state of ionisation and excitation. This depends primarily on the temperature of the atmosphere of the star and to a much lesser extent on its density, or equivalently pressure. The original sequence of spectral types, which was labelled ABC . . ., was then found to make much more sense as a temperature sequence if it was reordered and at the same time some of the classes were found to be superfluous. The current spectral classification in order of decreasing surface temperature is OBAFGKMRNS*. The main characteristics of the spectral types are shown in Table 1 and the approximate effective temperature corresponding to the different spectral classes is shown in Table 2. Spectral types OBA are called early spectral types and G–S late spectral types.

As mentioned above the density of the stellar atmosphere has some influence on the spectrum. There are both *K giants* and *K dwarfs* (stars on the lower main sequence are

Table 1. Main features in the spectrum of different spectral types

O:	Ionised helium and metals, weak hydrogen
B:	Neutral helium, ionised metals, hydrogen stronger
A:	Balmer lines of hydrogen dominate, singly ionised metals
F:	Hydrogen weaker, neutral and singly ionised metals
G:	Singly ionised calcium most prominent, hydrogen weaker, neutral metals
K:	Neutral metals, molecular bands appearing
M:	Titanium oxide dominant, neutral metals
R,N:	CN, CH, neutral metals
S:	Zirconium oxide, neutral metals.

* The usual way of remembering the present order is through the mnemonic 'Oh be a fine girl (guy) kiss me right now sweetheart'. The concept of thermodynamic equilibrium is discussed in Appendix 1.

frequently called dwarfs). The giants are much more luminous than the dwarfs but they have the same effective temperature. From equation (2.3), it is then clear that they must have very much larger radii. If their masses are not greatly different, and that proves to be the case, their surface densities and pressures must correspondingly be much lower. The level of ionisation at a given temperature is dependent on the pressure, as is discussed in Appendix 1. This means that the spectra of giants and dwarfs of the same surface temperature and chemical composition are not identical. This provides criteria for deciding whether a distant star of unknown distance is a giant or a dwarf. Once that is determined, a good estimate of its distance can be obtained.

The resulting luminosity criteria enable stars to be grouped into *luminosity classes*. These are labelled I, II, III, IV, and V. Class I contains the most luminous supergiants and class II the less luminous supergiants. Class III are the normal giants and Class IV the so-called subgiants which lie closer to the main sequence. Class V are main sequence stars and most stars fall into that class. The white dwarfs are not included in this classification. An astronomer wishing to characterise a star briefly will then describe it, for example, as a B2II star.

The specification of a spectral type and a luminosity class does not completely define the properties of a star. It may have variable light output, rotate rapidly or possess a strong magnetic field, be a member of a close binary or show activity in the sense that the Sun does through sunspots and flares. Similar stars are often grouped together.

There are thus cepheid variables, named after δ Cephei* the first such variable studied, and RR Lyrae variables. In this book I shall be concerned with several such groups of stars such as the pre-main-sequence T Tauri stars and the active RS Canum Venaticorum binary stars. These groupings may appear in some sense analogous to the classification of animals and plants into species but this analogy must not be pushed too far. There is no completely precise criterion for membership in a group and in addition the properties of a star change with time so that a T Tauri star may later become a completely normal main sequence star and even later a regular variable star. Never-

Table 2.   Effective temperature as a function of spectral type for main sequence stars

Spectral type	O	B0	A0	F0	G0	K0	M0	M5	R,N,S
T_e (K)	50 000	25 000	11 000	7 600	6 000	5 100	3 600	3 000	3 000

Each spectral type labelled by a capital letter is sub-divided into subclasses labelled by numbers, as in M5 above.

The Sun has type G2.

---

* Stars are labelled as members of particular constellations in the sky. Although it has long been realised that the constellations have no physical significance, it is a convenient way of indicating roughly where a star is to be found. For more accurate identification a star is given a *right ascension* and *declination* which are analogous to longitude and latitude on Earth. The brightest stars in a constellation are labelled α, β, γ ... and the fainter ones by numbers or letters. Doubled letters indicate variable stars, although the most luminous variables such as δ Cephei are not labelled in that manner.

theless, the identification of groups of stars with similar or related properties has been an important step towards an understanding of these properties.

Other gross stellar properties

As I shall explain in Chapter 3, from a theoretical point of view the three most important properties of a star are its mass, chemical composition and its age, the time since it was formed. The mass of the Sun can be obtained by a study of the orbits of the planets and the asteroids, which are governed by the gravitational attraction of the Sun. For other stars a direct measurement of mass is only possible if the star is a partner in a binary system which can be studied in great detail. Even then what is frequently determined is only a combination of mass and the inclination of the orbit to the line of sight. The stars with well-determined masses are those that have led to the relation which is shown in fig. 2. All other mass estimates are indirect and involve some input from theory. Although I have given an indication in discussing spectral classification that the chemical compositions of stars are not so variable that they dominate the differences in observed spectra, there are important composition differences from star to star particularly in the relative amounts of elements from carbon upwards in the periodic table relative to hydrogen and helium. The chemical composition of stars can only be obtained by a detailed discussion of how the emergent spectrum arises from the transport of radiation through the outer layers of the star. I shall comment further on this in discussing the Sun shortly.

The age of the Sun is believed to be known to some accuracy as I have already mentioned in Chapter 1 and I shall discuss this in greater detail in Chapter 8. It is not possible to obtain a value for the age of any other individual star but, as I shall explain in Chapter 6, estimates can be made of the ages of star clusters. These physical groupings of stars, either *globular clusters* or *open clusters* are believed to have originated together out of one mass of gas and the stars can be supposed to a close approximation to have the same chemical composition and age. It is possible to use the observed shape of the HR diagrams of star clusters and theoretical studies of stellar evolution to obtain ages for star clusters. This is discussed in *The Stars* and will be described briefly in Chapter 6.

Stellar rotation

Stellar spectra can also be used to obtain information about stellar properties other than surface temperature and chemical composition. Here an immediate distinction must be made between what is possible for the Sun and for other stars. The Sun provides a disk of light to the observer whereas any other star only appears as a point source of light. Observations can be made of individual surface areas of the Sun, and this will be described in detail later in this chapter, but the whole disk of any other star must be studied. There are only limited possibilities to obtain any spatial resolution of surface properties of stars, which I shall describe shortly.

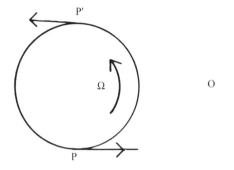

Fig. 3. As a star rotates with angular velocity Ω, an observer O receives blueshifted radiation from P and redshifted radiation from P'.

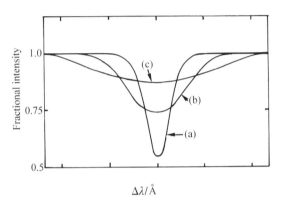

Fig. 4. The effect of rotation on the shape of a spectral line. (a) is the shape of the spectral line, in a non-rotating star, (b), in a star of moderate rotation and (c), in a rapidly rotating star.

What information can be obtained from the study of a whole stellar disk? All stars rotate and the radiation coming from the approaching and receding edges of a star is blueshifted and redshifted by the Doppler effect. This is illustrated in fig. 3. The effect on an individual spectral line is to broaden it, as shown in fig. 4. Provided this broadening is greater than any intrinsic sources of broadening at individual points on the star's surface, the rotation of the star can be detected; intrinsic broadening includes, for example, thermal Doppler broadening which arises because individual atoms have velocities towards or away from the observer determined by their temperature. What can be measured is $v\sin i$, where v is the rotation velocity at the star's equator and i is the angle of inclination of the axis of rotation to the line of sight. In general i is not known and only statistical estimates of the true value of v can be obtained by studying groups of otherwise similar stars for which the values of i may plausibly be supposed to arise from random orientation.

Stellar magnetic fields

A second property which can be deduced from a careful study of stellar spectra is the presence of a magnetic field in the stellar atmosphere. In the presence of a magnetic field an individual spectral line splits into two or more components through what is known as the *Zeeman effect* (see Appendix 2). The problem with the Zeeman effect is that the splitting depends on both the strength of the magnetic field and its orientation to the line of sight. Even if the magnetic field has a simple geometrical structure such as a dipolar field, the strength and orientation both vary across the stellar disk and the observed spectral line shapes and splitting arise from a superposition of a continuously varying set of line profiles. In particular, as a result of both intrinsic broadening and rotational broadening, the different Zeeman components are usually included in one broadened line instead of appearing as distinct lines. What can be determined is some average value of the surface field. As I shall explain later, this average dipole field for the Sun is very weak but the solar surface is covered with local regions of very much stronger field, with the flux which leaves the surface at one point returning at another. No useful information about solar magnetism could be obtained if the Sun were at a typical nearby stellar distance from us. Some stars are observed to have strong average surface fields ($\sim 1\,$T), particularly the so-called peculiar A stars (Ap stars). There are even stronger fields in some white dwarfs ($> 10^2\,$T) and neutron stars ($\sim 10^8\,$T).

Binary stars

Many if not most stars are members of binary or multiple systems. Some binaries are wide binaries in the sense that both of the components appear separately in a photograph (obviously this depends on the distance of the system), but the binaries with which I shall be concerned in this book are close spectroscopic binaries, where the presence of two stars only becomes apparent from the study of the spectrum of an apparently single star. If binary stars are sufficiently close, their gravitational interactions will be efficient enough to cause one or both of them always to show the same face to the other (in the same way that the Moon always shows the same face to the Earth). Because the orbital period of a close binary is very short, the individual stars are rapidly rotating. As I shall explain later, stellar activity, which is a principal concern of this book, is increased by rapid rotation and this means that the study of binary stars is very important. A further possibility is that in a close binary system there is an important linkage of magnetic flux from one star to the other.

Observations from space

So far I have been discussing observations made with optical telescopes on the Earth's surface. The ability to put telescopes into space and to make observations in regions of the electromagnetic spectrum where radiation is mainly absorbed before it reaches the Earth's surface, has made a substantial difference to our knowledge of other stars. Such spectral regions include the ultraviolet and X-ray regions and particularly important results have been obtained by the *International Ultraviolet Explorer* (IUE), which commenced observations in 1978 and which is still operational at the time of writing

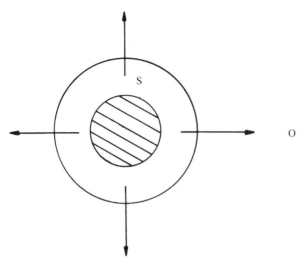

Fig. 5. Mass flows outwards as shown from the surface S of a star. Because the material being lost is at least partially transparent to radiation, an observer at O sees very wide spectral lines produced by material in both hemispheres of the star.

but is now expected to cease operation in the near future. Two important results which have been found through use of IUE and other telescopes refer to *stellar mass loss* and *stellar coronae*. Mass loss can be deduced from the observations of very broad spectral lines indicating the existence of velocities in excess of the escape velocity from the star in the star's atmosphere (fig. 5). The broad spectral lines which are produced by matter flowing out from both faces of the star could in principle also be produced by matter falling into the star but in most cases it is clear that the only plausible explanation is that matter is being expelled; there is no source of external matter to maintain an inflow for an astronomical timescale. The opposite is, of course the case in star formation when the star is being assembled. It has become clear that mass loss is common for stars in many parts of the HR diagram and that the final mass of a star may be very much less than its initial mass.

I shall discuss the solar corona later in this chapter. It is the high solar atmosphere and it is very much hotter than the solar surface with the matter having a kinetic temperature (a measure of the mean kinetic energy of the particles) of 1–2×10^6 K. The mass of the corona is very small compared to the mass of the Sun but somehow the coronal matter has to acquire an undue share of energy. Ultraviolet and X-ray observations have shown that high temperature coronae are common in late-type stars.

Detection of the large scale surface features

In this chapter I am describing how information has been obtained about the average properties of stars other than the Sun. I shall leave until Chapter 6 any discussion of irregular and violent activity in stars apart from making a few remarks about how it is possible to get some spatial resolution of stellar properties, even though a stellar disk cannot be observed. Such resolution only refers to rather large portions of a stellar disk and cannot approach the resolution possible on the Sun. The two techniques involve stellar rotation and the use of eclipses or occultations.

If a star's properties vary significantly over its surface, the integrated properties of the visible disk may vary as the star rotates. Two examples are the following. In Ap stars the measured strength of the surface magnetic field varies with time. This can be understood in terms of an oblique rotator model in which a regular magnetic structure fixed in the stellar surface is rotating about an axis which makes an angle with its own axis of symmetry, or near-symmetry. In the simplest example of a dipolar field with its centre at the stellar centre, the variations should be symmetrical about the mean value. Observed variations frequently do not have this property and an attempt can then be made to fit them with more complicated magnetic field geometries of which the simplest is an off-centred dipole. As I shall explain in Chapter 6, some stars possess spots which are very much larger than sunspots. They are cooler than the remainder of the surface of the star and as a result the observed luminosity of the star varies as the spot appears and disappears. Of course the observation here is the luminosity variation, which is interpreted as being due to a spot. The stellar rotation can only provide spatial information about features which last for one or many rotation periods.

By an eclipse I mean a star disappearing behind its companion in a binary system. Occultation usually refers to lunar occultations when a star apparently passes behind the Moon. If the radiation from the disk of the star was reaching us equally from all parts of the disk, the apparent luminosity of the star would be reduced by occultation or eclipse in proportion to the invisible region of the disk. In fact this is never the case because of the phenomenon known as limb darkening. Radiation which reaches us from the centre of the disk leaves the star at right angles to its surface and it is more intense than radiation from the edge (limb) of the disk, which moves tangential to the stellar surface. Observations of eclipses and occultations can be used to test theoretical predictions of what the limb darkening should be. They might also in principle be used to detect anomalously dark or bright surface features.

This completes my general introduction to the properties of stars other than the Sun and I now turn to the Sun itself for the remainder of this chapter.

The gross properties of the Sun

The Sun is a G2V star and must be considered typical of such stars in the disk of our Galaxy, which have a generally similar composition to it, unless something causes us to change this view. Its main overall properties are

$$M_\odot = 1.99 \times 10^{30}\,\text{kg,}$$
$$L_\odot = 3.83 \times 10^{26}\,\text{W,}$$
$$r_\odot = 6.96 \times 10^{8}\,\text{m,}$$
$$T_{e\odot} = 5780\,\text{K.}$$

(2.5)

The mass of the Sun and the mean distance of the Earth from the Sun are both determined by an application of Kepler's laws to the observed motions of bodies in the solar system. The use of radar has greatly increased the accuracy of some measured distances in the solar system. A detailed discussion of what is involved is extremely complicated and I shall therefore not attempt it. To be accurate what is obtained is not $M_\odot$ but $GM_\odot$ and a separate terrestrial determination of G is required to give $M_\odot$. G is by far the least well known of the fundamental physical constants but it provides a solar mass to better than the accuracy which we require. Once the distance of the Earth to the Sun is known, the observed angular diameter of the Sun can be converted to a linear diameter or radius.

The solar luminosity

The solar luminosity shown in (2.5) is the bolometric luminosity, which is what has to be compared with predictions of theoretical models of the Sun to be discussed in the next Chapter. It is usually described in terms of the *solar constant* which is the energy crossing unit area of the Earth's surface perpendicular to the direction from the Earth to the Sun in each second. Because the Earth's orbit about the Sun is not quite circular, a quoted value must refer to a particular orbital position and distance from the Sun. In SI units it is in Wm^{-2}. The solar constant cannot be measured directly on the Earth's surface because, although the Earth's atmosphere is effectively transparent to a large fraction of the solar radiation, some radiation in both the ultraviolet and the infrared is strongly absorbed by the atmosphere. Before space astronomy was possible, the amount of missing radiation in these frequencies had to be estimated. As a result there was an uncertainty in the solar constant and in the resulting solar luminosity. The quoted value of the solar luminosity has been modified as a result of measurements across the whole frequency range. The value quoted above differs slightly from that used in *The Stars*.

The major result of studies of the solar luminosity in recent years has been the discovery by instruments on two different spacecraft that the solar constant is not constant but that it is variable by a fraction of one per cent. The solar luminosity has been monitored by two radiometers on spacecraft, ACRIM on *SMM* and ERB on *Nimbus* 7. Some results from both spacecraft are shown in figs. 6 and 7. ACRIM results have shown that the presence of several large sunspots, which are cooler than their surroundings and radiate less energy, depresses the solar luminosity by ~ 0.1 per cent. This shows that the heat blocked by sunspots cannot be re-radiated at once. I shall discuss sunspots and the solar cycle later in the chapter. Here I will simply say that, when

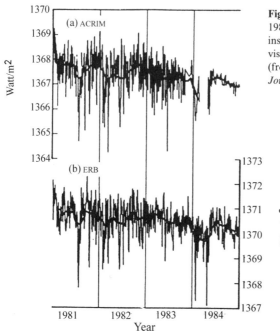

Fig. 6. Solar irradiance at Earth for the period 1981 to 1984 measured by the ACRIM and ERB instruments. In each case a long-term trend is visible although there are short period oscillations (from Foukal, P. & Lean, J, *Astrophysical Journal*, **328**, 347, 1988).

the radiometer measurements are followed throughout a solar cycle, it is found that the solar luminosity varies, so that it is highest when the number of sunspots is a maximum. This is opposite to the naive expectation that the luminosity would be reduced by a large sunspot area. Although sunspots do reduce the flux of radiation in their immediate region, over the whole solar surface there is an increase in radiation which more than makes up for the local decrease. I shall comment further on this in later chapters. Meanwhile it is enough to say that the solar luminosity is almost exactly constant and it can be regarded as such in discussions of average solar properties. Once the solar luminosity is known its effective temperature follows from equation (2.3).

The solar age

The age of the Sun is believed to be about 4.6×10^9 yr. I shall discuss how this age is obtained in Chapter 8. All that I need to say at the moment is that studies of the abundances of heavy radioactive elements and their decay products enable the time to be found since meteorites and lunar and terrestrial rocks solidified. Because of internal activity in the Earth, which still continues, and the Moon, lunar and terrestrial rocks may be younger than the Earth and the Moon. There should not have been any such activity in meteorites once they formed and many of the meteorites have an age of about 4.55×10^9 yr, which is also approximately the age of the oldest deposits found on Earth. If it is supposed that, as most astronomers do, the solar system formed as one unit, the

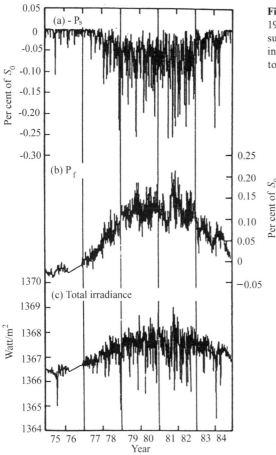

Fig. 7. Solar irradiance at Earth for the period 1975 to 1984. The reduction in irradiance due to sunspots (a) is more than counterbalanced by the increase due to bright faculae (b) leading to the total irradiance variation (c) (source as fig. 6).

age of the Sun should be very close to that of the meteorites, giving the approximate figure of 4.6×10^9 yr. There are some suggestions that the planets might have been formed later than the Sun which would allow the Sun to be older than 4.6×10^9 yr. Such theories are currently out of favour as I shall explain in Chapter 8. The theory of that type, which has most current support, requires the Sun to have almost the same age as the rest of the solar system. This would not require any modification to our age estimate.

The solar composition

In addition to the mass, the solar chemical composition is the major input to theoretical models of the Sun. Only the surface layers of the Sun, or any other star, can be studied. I shall comment in the next chapter on why it seems likely that the surface chemical composition of the Sun, and the other stars, is similar to the initial chemical composition. Here I will give a brief account of how the chemical composition of the Sun is

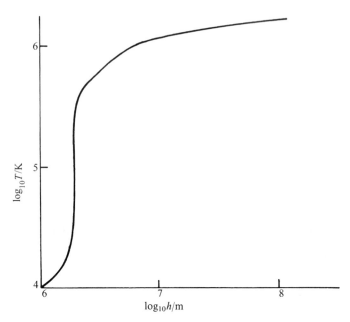

Fig. 8.   Kinetic temperature as a function of height above the solar photosphere. The temperature first rises slowly in the chromosphere, followed by a very rapid rise in the transition region before reaching a value in excess of 10^6K in the corona. What is shown is an average profile as there are very large variations over the solar surface.

determined. All the detailed results refer to studies of absorption spectral lines from radiation leaving the visible surface, *photosphere*, of the Sun. Above the photosphere are the *chromosphere, transition region* and *corona* (fig. 8). Spectral lines produced here are mainly emission lines and they are not so useful for obtaining accurate element abundances as I shall explain later.

Spectral line formation

A very approximate explanation of the formation of absorption lines is as follows. In the deep interior of any star conditions are very close to thermodynamic equilibrium. This means that radiation essentially possesses the Planck distribution (2.2) and flows equally in all directions. There is a very small departure from isotropy which provides the outward flow of radiation. As the surface of the star is approached and the temperature drops, the isotropic radiation flux represented by the Planck function decreases greatly in magnitude, so that the relative departure from isotropy must be greater if the outward flow of radiation is to be maintained. In the deep interior, photons which are emitted, will be absorbed after travelling a minute fraction of the stellar radius. Near the surface, this is no longer true and the photons have a finite probability of escaping from the star.

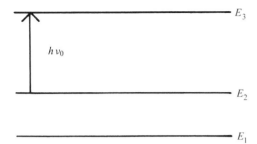

Fig. 9. Formation of an absorption line. An absorption line at frequency ν_0 is formed if electrons move from atomic energy level of energy E_2 to the higher level of energy E_3 where $h\nu_0 = E_3 - E_2$.

The visible surface is roughly speaking the level at which photons have a 50 per cent chance of being absorbed before they leave the star. This level is not independent of frequency. If absorption is particularly strong at a frequency ν_0, because some element has two energy levels giving a spectral line there (fig. 9), the level from which radiation at frequency ν_0 can escape will be further from the centre of the star and at a lower temperature. As a result the radiation at that frequency will be less intense than at the neighbouring frequencies and an absorption line will result.

The occurrence and strength of a spectral line now depends on three factors:

(i)   the amount of the chemical element in unit mass in the atmosphere of the star;
(ii) the probability that the element will be the lower energy state which can lead to the transition;
(iii) the probability that a photon of frequency ν_0 will be absorbed.

The third factor is essentially straightforward physics but for most atoms it is not possible to calculate the probabilities and their measurement requires extremely precise experimentation. If I assume that this is known, (i) can only be obtained from an observation of the strength of the spectral line if (ii) is also known. However (ii) raises considerable problems.

If conditions are close to thermodynamic equilibrium, the state of ionisation and excitation of any atom depends only on density, temperature and chemical composition as is explained in Appendix 1. In thermodynamic equilibrium the number of ionisations s^{-1} is equal to the number of recombinations. The latter is proportional to the number of free electrons because these initiate the recombinations. Thus the dependence of ionisation of any element on chemical composition arises through the total number of electrons present in the atmosphere, and these may be released by the ionisation of any other element as well as the element being studied. This sounds very complicated but, as the electrons will effectively only be provided by the more abundant elements, the complication is not quite as significant as it sounds. It is however obvious that conditions in the atmosphere of a star differ from those of thermodynamic equilibrium, in particular because the radiation does not flow equally in all directions. The first approach to the theoretical treatment of spectral lines was to assume that the atmosphere was in a state known as *local thermodynamic equilibrium* (LTE). In LTE all of the conditions of true thermodynamic equilibrium are supposed to apply except that the radiation is not given by the Planck function. If these assumptions are made, it is possible to find for what abundance of the specified chemical element the theoretically calculated spectral line has the best fit to the observed spectral line.

However the abundances found in this manner are no better than the assumptions which have been used in obtaining them. In principle it is necessary to consider all of the different processes, collisional or radiative, which populate or depopulate the different states of ionisation and excitation to discover whether or not the populations are close to the LTE values. If they are not a non-LTE approach, which is much more complicated, must be used. The purpose of this brief discussion has not been to provide an understanding of the manner in which abundances can be obtained from a study of spectral lines but to make it clear why even now the chemical composition of the solar atmosphere cannot be regarded as precisely known. Note however that the main problem is that of obtaining the absolute abundances in one star such as the Sun. It is much easier to estimate the relative abundances of elements in the Sun and another G2V star. If, for example, iron spectral lines are studied in two such stars, their relative strengths should depend only on the different element abundances because the overall properties of the two atmospheres should be essentially the same.

The emission lines from the chromosphere and corona are formed where conditions are certainly very far from thermodynamic equilibrium. It is as a result much more difficult to estimate the populations of the upper levels of ions from which the emission lines arise. This is particularly the case when the lines come from an element in a high state of ionisation. As a result abundances determined from these emission lines are not usually thought to be as accurate as abundances determined from photospheric absorption lines.

Additional problems in determining the solar composition

The abundance of any element can only be found if it has absorption lines in the region of the solar spectrum which is accessible to study. There are many elements for which visible absorption lines do not appear in the solar spectrum because the energy levels which might lead to such lines are not populated at the temperature of the solar atmosphere. In most cases these elements do not play a major part in the composition of the Sun. As I shall explain in Chapter 8, for those elements which can be studied both in the Sun and in the Earth and meteorites, relative abundances are very similar and this gives support to the idea that the whole solar system had a common origin. It therefore appears appropriate to use for the Sun abundances deduced from terrestrial and meteoritic values. There is however one major exception to this. The Earth and meteorites are deficient in volatile gases which will have escaped during their formation; thus hydrogen, nitrogen and oxygen are all underabundant in the Earth relative to the Sun. The same should also be true of helium but unfortunately helium does not figure in the visible spectrum of the solar photosphere. At the photospheric temperature all of the helium is in its ground state and all spectral lines arising from the helium ground state are well into the ultraviolet, where the Sun emits very little radiation. There is therefore no directly observed photospheric abundance of helium.

Helium in the Sun

It is perhaps paradoxical that helium, which was discovered on the Sun before it was found on Earth, should have an uncertain abundance. Sir Norman Lockyer discovered an unknown element in the solar spectrum in the 1860s. He named the element helium after the Greek Sun god *Helios*, but it was only discovered on Earth by Ramsay in 1895. Lockyer's observations were of an emission line in the solar chromosphere, which was detected during a solar eclipse. Some attempts to determine a solar helium abundance can be made by a study of the chromosphere, corona, solar wind and energetic particles produced in solar flares and I shall mention the latter later. It is, however, generally believed that comparison of theoretical models of the Sun with a variety of observations, which I shall mention in Chapter 3, is more likely to provide a reliable value for the solar helium abundance. The inability to obtain a direct observation of photospheric helium abundance is common to all stars with a relatively low surface temperature. This includes the low mass unevolved stars in globular star clusters, which are the oldest known stars in the Galaxy.

It is very important to obtain a good value for such stellar helium abundances. The currently favoured *hot big bang cosmological theory* predicts that before galaxies formed the chemical composition of the Universe should have been a mixture of hydrogen and helium, with the helium contributing about 23 to 24 per cent by mass. The oldest globular cluster stars should have a helium abundance close to that, while the Sun, which may have been formed when the Galaxy was up to 10^{10} yr old, should contain rather more helium. The additional helium along with the heavier elements in the Sun will then have been produced in early generations of stars in the Galaxy. A figure of 27 to 28 per cent for the solar helium abundance is generally considered reasonable, although some workers deduce a value much closer to the assumed primeval abundance.

A recent compilation of the relative abundances of different elements in the solar atmosphere is shown in Table 3.

The solar granulation

I now turn to some properties of the Sun which are variable in space and time on a short timescale but which are not concerned with the magnetically active Sun. As I shall explain in the next chapter there is a layer near the solar surface in which the principal mechanism of energy transport is convection. This means that elements of matter are moving upwards and downwards in the solar atmosphere. Here I give a brief description of the observational evidence for the convective motions, the solar *granulation*.

It has been known for a long time that the solar surface is not uniformly bright. It was believed that this was associated with the convection, with the rising elements of material being hotter and hence brighter than the falling elements. The available resolution of ground-based telescopic observations was not good enough to see the detailed structure. Such resolution was achieved by the balloon-borne solar telescope, *Stratoscope*, in the

Table 3.   Logarithmic elemental abundances in the Sun

Element	Abundance	Element	Abundance	Element	Abundance
^{1}H	12.00	^{25}Mn	5.39 ± 0.03	^{57}La	1.22 ± 0.09
^{2}He	[10.99 ± 0.035]	^{26}Fe	7.50 ± 0.04	^{58}Ce	1.55 ± 0.20
^{3}Li	1.16 ± 0.10	^{27}Co	4.92 ± 0.04	^{59}Pr	0.71 ± 0.08
^{4}Be	1.15 ± 0.10	^{28}Ni	6.25 ± 0.04	^{60}Nd	1.50 ± 0.06
^{5}B	(2.6 ± 0.3)	^{29}Cu	4.21 ± 0.04	^{62}Sm	1.01 ± 0.06
^{6}C	8.55 ± 0.05	^{30}Zn	4.60 ± 0.08	^{63}Eu	0.51 ± 0.08
^{7}N	7.97 ± 0.07	^{31}Ga	2.88 ± (0.10)	^{64}Gd	1.12 ± 0.04
^{8}O	8.87 ± 0.07	^{32}Ge	3.41 ± 0.14	^{65}Tb	(−0.1 ± 0.3)
^{9}F	[4.56 ± 0.3]	^{37}Rb	2.60 ± (0.15)	^{66}Dy	1.14 ± 0.08
^{10}Ne	[8.08 ± 0.06]	^{38}Sr	2.90 ± 0.06	^{67}Ho	(0.26 ± 0.16)
^{11}Na	6.33 ± 0.03	^{39}Y	2.24 ± 0.03	^{68}Er	0.93 ± 0.06
^{12}Mg	7.58 ± 0.05	^{40}Zr	2.60 ± 0.03	^{69}Tm	(0.00 ± 0.15)
^{13}Al	6.47 ± 0.07	^{41}Nb	1.42 ± 0.06	^{70}Yb	1.08 ± (0.15)
^{14}Si	7.55 ± 0.05	^{42}Mo	1.92 ± 0.05	^{71}Lu	(0.76 ± 0.30)
^{15}P	5.45 ± (0.04)	^{44}Ru	1.84 ± 0.07	^{72}Hf	0.88 ± (0.08)
^{16}S	7.21 ± 0.06	^{45}Rh	1.12 ± 0.12	^{74}W	(1.11 ± 0.15)
^{17}Cl	[5.5 ± 0.3]	^{46}Pd	1.69 ± 0.04	^{76}Os	1.45 ± 0.10
^{18}Ar	6.52 ± 0.10]	^{47}Ag	(0.94 ± 0.25)	^{77}Ir	1.35 ± (0.10)
^{19}K	5.12 ± 0.13	^{48}Cd	(1.77 ± 0.11)	^{78}Pt	1.8 ± 0.3
^{20}Ca	6.36 ± 0.02	^{49}In	(1.66 ± 0.15)	^{79}Au	(1.01 ± 0.15)
^{21}Sc	3.17 ± 0.10	^{50}Sn	2.0 ± (0.3)	^{81}Tl	(0.9 ± 0.2)
^{22}Ti	5.02 ± 0.06	^{51}Sb	1.0 ± (0.3)	^{82}Pb	(1.95 ± 0.08)
^{23}V	4.00 ± 0.02	^{56}Ba	2.13 ± 0.05	^{92}U	(< −0.47)
^{24}Cr	5.67 ± 0.03				

The logarithm to base 10 of the number of atoms of each element is shown, with uncertainties in the value, on a scale in which the number of hydrogen atoms is 10^{12}. Values in parentheses are uncertain. Values in brackets are not obtained from photospheric spectra but are based on other solar or astronomical data. (The values shown are courtesy of N Grevesse and A Noels.) No direct abundance estimate is possible for those elements not shown.

1960s. It is now known that the solar surface is covered with a polygonal structure, which consists of bright granules with narrow dark lanes in between. There is thus a pattern of rising cells with falling matter at the boundaries. The pattern has a general similarity to a hexagonal pattern known as Bénard convection which is observed in laboratory experiments. A typical velocity is $2\,\text{kms}^{-1}$, which can be measured directly through the Doppler effect. A granular scale of about 700 km is typical with temperature differences of order 100 K between dark and bright regions. Individual granules last about 10 to 20 minutes; sometimes they disappear quietly but sometimes they explode. Outside the regions occupied by sunspots the granulation is very regular.

There is also a larger scale smaller amplitude motion known as the *supergranulation*. This has a typical velocity of $0.3 \, \mathrm{kms}^{-1}$ and a length scale of about $3.5 \times 10^4 \, \mathrm{km}$. This therefore represents a very much larger scale and deeper sample of the convection zone than the granulation, which can in some sense be regarded as the froth on the surface. Even so, this scale is in turn less than the depth of the entire convection zone, which, as I shall describe in Chapter 3, is believed to occupy the outer $2 \times 10^5 \, \mathrm{km}$ of the Sun or not far short of a third of the solar radius. Although the convection zone occupies a significant fraction of the volume of the Sun it does not contain much of the mass because the central density is very much higher than the surface density. The magnetic field of the Sun outside sunspots tends to be concentrated at supergranular boundaries.

Solar oscillations

A surprise discovery in the 1960s was the *five-minute oscillations* of the solar surface. The amplitude of these vertical oscillations is very variable but the period is of order five minutes. It was eventually realised that the observations could be understood in terms of a superposition of many normal modes of oscillation of the Sun. As an understanding of these oscillations requires a significant theoretical input I will defer any further discussion to Chapter 3.

The granulation and the solar oscillations represent time-variable properties of the *quiet Sun*. I now turn to a brief survey of observations of the *active Sun*. This will include a discussion of the corona and the solar wind because, although they are also attributes of the quiet Sun, they are significantly affected by magnetic activity.

Sunspots

The first realisation that the properties of the Sun are not the same at all times came with the discovery of sunspots, small dark regions which disturbed the smooth brightness of the solar surface and which appeared and disappeared. Although sunspots had certainly been seen before the early years of the seventeenth century, particularly in China, they were discovered in the West shortly after the invention of the telescope. They were one of Galileo's pieces of evidence against the medieval world picture in which everything beyond the Moon was supposed to be pure and unchanging. They were also observed in the same year, 1611, by Fabricius and Scheiner. In these early studies it was already clear that a sunspot consisted of a central dark region known as the *umbra* and a surrounding less dark region known as the *penumbra*.

Sunspots were immediately seen to drift across the surface of the Sun. This could in principle be the motion of dark clouds across the bright visible surface of a non-rotating Sun. Alternatively the sunspots could be fixed on the surface and be carried around by the rotating Sun. This was soon accepted as the explanation and it could then be seen that sunspots could survive for one or more rotation periods; the rotation of the Sun can of course now be studied from the Doppler shift of spectral lines received from different parts of the solar surface although this observation is complicated by the various photospheric motions, which have already been mentioned. The observations of sunspots showed that the rotation of the Sun was not the same at all latitudes. There is an *equatorial acceleration* giving an equatorial rotation period of 27.7 days compared to 28.6 days, at latitude 40°. Another early discovery was that of the *Wilson depression*. The surface of a sunspot is depressed below the surface of the surrounding bright Sun, which means that the spot is certainly not a dark cloud floating on a smooth bright surface. However even at the time that the Wilson depression was discovered, Sir William Herschel was speculating that the Sun might be inhabited, with the sunspots being dark clouds in the atmosphere of a habitable globe.

Sunspot cycle, butterfly diagram, Maunder minimum

In the middle of the 19th century two further important discoveries were made. Schwabe showed that the number of sunspots was approximately periodic with a period of about 11 years. Some data relating to this *sunspot cycle* are shown in fig. 10. Carrington then showed that during any sunspot cycle the places at which sunspots were principally seen migrated in latitude, with spots being at higher latitudes early in the cycle. This gives rise to the *butterfly diagram* shown in fig. 11. It can be seen in this figure that sunspots are essentially a low latitude phenomenon. There are very few spots further than 30° from

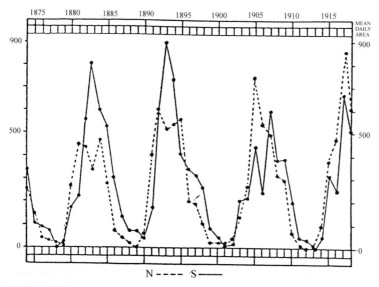

Fig. 10. Total sunspot numbers for the two solar hemispheres as a function of time from Maunder (*Monthly Notices Royal Astronomical Society*, **82**, 534, 1922).

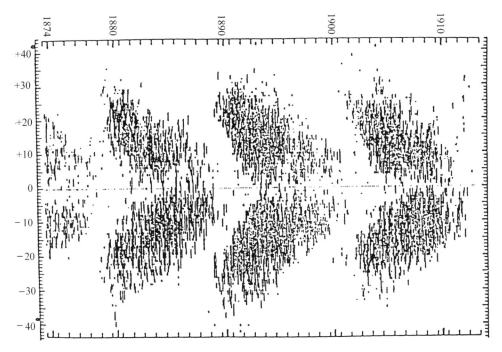

Fig. 11. A butterfly diagram showing migration of sunspots in latitude for three solar cycles (source as fig. 10).

the solar equator. Later in the book we shall see that this is not necessarily the case for other active stars.

Although sunspots were discovered in 1611, they were not studied systematically until the establishment of observatories such as the Paris Observatory and the Royal Greenwich Observatory in the second half of the 17th century. Even then the discoveries of the periodicities in sunspot properties were hindered by there being extremely few sunspots at the time when serious telescopic study of the Sun was starting. This was realised by Maunder two hundred years later and the period of low sunspot number is now known as the *Maunder minimum*. The period of the Maunder minimum was one of very low temperatures, at least in the northern hemisphere, where they are good records; this was the so-called *little ice age*. As I mentioned there has been a small variation of solar luminosity during a recent solar cycle, with the lowest luminosity being when sunspot numbers were low. It is therefore not implausible that the solar luminosity was even lower during the Maunder minimum and that this affected the Earth's climate. In Chapter 5 I shall discuss this problem of a change in behaviour of the solar cycle and will provide evidence that such change had happened in the past, even though sunspot records are not available.

Now that I am discussing detailed variations from point to point on the solar surface, I should perhaps comment on what resolution is currently available. On the observed Sun near the centre of the disk an angular separation of one second of arc represents a distance of about 670 km. Current observational techniques enable a resolution of about

0.3″, or 200 km, if spectroscopy is not being used, and about 0.5″ using spectroscopy. Although these distances are very small compared to the solar diameter, they are not so small that one can be confident that one is able to study the finest scale phenomena on the solar surface. There is however no real problem with sunspots which have observed radii in the range of about 1800 km to 2500 km; a few even larger spots have been observed. Spots are also generally quite symmetrical. Even at sunspot maximum less than one per cent of the Sun is spotted.

Sunspots are darker than their surroundings, which means that they are radiating less energy per unit area and in turn that they have a lower temperature. A characteristic umbral temperature of about 4000 K can account for the difference between umbral and normal photospheric brightness. This raises two immediate questions. Why is the spot cooler than its surroundings and what happens to the radiant flux which is unable to escape through the spot? I shall return to these two questions in Chapter 5. The penumbra has a temperature intermediate between the umbra and the normal photosphere and it also has a filamentary structure. Another important property relating to the penumbra is the *Evershed effect*. This is a radial outflow of matter with the velocity increasing outwards, with a characteristic speed of 1 to 2 kms^{-1}. One clue relating to different properties in a sunspot and its surroundings comes from studies of the solar granulation, which I mentioned earlier. This rather regular cellular pattern which is present in the normal photosphere is absent in the spot.

Magnetic field of sunspots

The key observation was the discovery by Hale in 1908 of the *Zeeman splitting* of spectral lines. In the absence of a magnetic field several quantum mechanical states of an atom may possess the same energy, even though they are characterised by different quantum numbers. A magnetic field destroys the symmetry of the system by introducing a preferred direction and there is a splitting in the energy levels. If a spectral line is now observed which arises from a transition involving the states concerned, it is split into several components. This is illustrated schematically in fig. 12 and the Zeeman effect is discussed further in Appendix 2.

Hale's observations showed that spots usually but not always occurred in bipolar pairs. Furthermore it soon became clear that the magnetic polarity of the leading spot in the pairs, in terms of the direction of solar rotation, changes from one 11 year cycle to the next. Thus, in reality, there is a 22 year magnetic cycle. The magnetic properties of a bipolar pair can now be understood in the manner shown in fig. 13. A tube of magnetic flux, which is anchored lower down in the outer envelope of the Sun, has broken through the solar surface and spots are observed where it intersects the photosphere. The appearance of sunspot pairs is then associated with flux rising from the interior. It is then clear that the production and motion of such magnetic flux tubes is crucial to an understanding of the solar cycle. The magnitude of the magnetic induction in a sunspot umbra is typically about 0.3 T and it falls to between 0.15 T and 0.1 T in the penumbra.

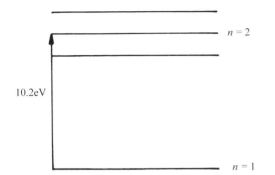

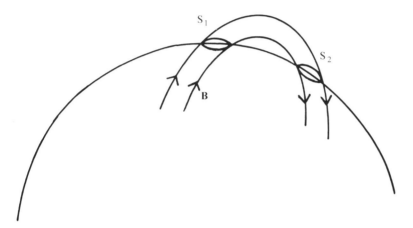

Fig. 12. The ground state and first excited state of the hydrogen atom, showing that the first excited state is split into three energy levels in the presence of a magnetic field (not to scale).

Fig. 13. A magnetic flux tube breaks through the solar surface producing a sunspot S_1 and returns through a second spot, with opposite polarity, S_2.

The magnetic field is approximately vertical in the umbra and it has an increasing inclination to the vertical through the penumbra. As I shall explain in Chapter 5 it is believed that the low temperature in a spot is produced by the magnetic field. Energy transport just below the photosphere is largely by convection and a magnetic field inhibits convection, thereby cooling the spot. The Wilson depression is also a consequence of the strong magnetic pressure inside the spot.

Magnetic fields outside sunspots

Hale's early magnetic observations suggested that the Sun had an overall dipolar magnetic field, which would have a typical strength near the Sun of about 5×10^{-3} T. Near the surface such a field would be swamped locally by the very much stronger sunspot fields, but at a large distance from the Sun it would appear as an object with such a dipolar field. It subsequently became clear that Hale's interpretation of his observations was erroneous and any general dipolar field is now known to have a strength more like 10^{-4} T. However, this very weak dipolar field is observed to reverse with the magnetic cycle. When I discuss the solar corona and solar wind it will become clear that in no sense does the Sun now appear as a dipole sitting in empty space.

The field outside sunspots was once believed to be characteristic of this very weak dipolar field, but it is now known that this picture is completely false. Almost all of the photospheric field which is not in sunspots is concentrated in small magnetic elements with magnetic induction typically between 0.1 and 0.15 T. This means that there must be a tendency in the convection zone to concentrate magnetic flux into ropes or filaments, whether they be the massive ones in large sunspots or the very small photospheric filaments.

Solar eclipses, chromosphere, corona

Many important discoveries relating to the outer solar atmosphere were made during a total solar eclipse. It is a remarkable coincidence that the Earth should have a satellite whose size and distance are such that it is possible for the Moon to produce a total eclipse of the Sun; it is even more remarkable that the apparent sizes of the Sun and Moon are so nearly the same that total eclipses of the Sun can just occur. During a solar eclipse the light of the photosphere is cut out and it is possible to observe the much lower intensity emission from the *chromosphere* and the *corona*.

Fig. 14. A photograph of the solar corona taken at the eclipse of 1966 November 12 by Newkirk, G A, Jr. (Royal Astronomical Society.)

Visually it is the corona which is apparent as it extends a large distance above the visible surface of the Sun with arcs and streamers shown in fig. 14. It became apparent quite early that the shape of the corona is different at sunspot maximum and sunspot minimum with C A Young saying in 1878, 'The eclipse of 1878 has demonstrated that the unknown cause whatever it may be which produces the periodical sunspots at intervals of about eleven years also affects the coronal atmosphere of the Sun.'

The visible appearance of the corona includes both open streamers and closed loops. It is now known that these are associated with magnetic field lines which rise from the visible surface of the Sun. Those which return to the surface provide the closed loops, whereas the open streamers are related to field lines which extend a large distance from the Sun. As I shall explain shortly they also carry the *solar wind*, which is a continuous loss of mass from the Sun. The proportion of closed loops to open field lines varies during the solar cycle.

When the light from the solar corona was studied spectroscopically it provided a great puzzle because many of the strongest spectral lines could not be identified as being due to any known chemical element. The earlier discovery of helium through an unidentified line in the spectrum of the solar chromosphere many years before its identification on Earth led to the idea that the coronal lines were created by another new element which was named *coronium*. At around the same time spectral lines in gaseous nebulae were attributed to a further element called *nebulium*. It was of course not obviously true that all of the lines in either case were produced by one element or that coronium and nebulium were distinct. A serious difficulty with these proposals became clear when there was no remaining space in the periodic table of the elements for either nebulium or coronium.

Eventually it was realised that both the coronium and nebulium lines were produced by well-known elements but that they were produced by very unusual transitions which could not be observed in the laboratory. These *forbidden* spectral lines arise from a transition in which an electron can spend an unusually long time in an excited state before it returns to a lower level. In laboratory conditions the atom will undergo many collisions in the time required for radiative decay to occur and the electron will either return to the ground state without emission of radiation or move to a higher level. As a result the forbidden spectral line will not be observed. In the corona and gaseous nebulae the density of matter is extremely low and collisions are sufficiently infrequent that the forbidden transition can be observed. The concept of forbidden transitions is discussed further in Appendix 3.

The problem with understanding the coronal spectrum was not just that the lines involve forbidden radiation but that they arise not from neutral atoms but from highly ionised atoms. This was a complete surprise at the time. The high degree of ionisation could only be present if the kinetic temperature of the corona was very high, of the order of 10^6 K or more. Kinetic temperature is a measure of the mean kinetic energy of the gas particles. The corona does not have a true thermodynamic temperature because the radiation present is mainly characteristic of the photospheric temperature. However, the radiation which is emitted by the coronal material is characteristic of a much higher kinetic temperature. With ground-based observations it is not possible to study most of the ultraviolet region of the spectrum or X-rays. Observations from space show that the corona emits in both of these spectral regions and this confirms the high temperature.

The very high degree of ionisation is illustrated in a list of coronal line emissions covering the wavelength range from 331.8 nm to 1079.8 nm published in 1971. This includes lines of Ca XII to Ca XV, Fe XI to Fe XV and Ni XIII to Ni XVI; here the Roman numeral is one more than the number of electrons removed from the atom, so that in the last case nickel has lost 15 of its 28 electrons. With the advent of ultraviolet astronomy from space further forbidden lines have been found in that spectral region, many of them corresponding to lower levels of ionisation than the visible lines. Not all coronal spectral lines are forbidden but the forbidden lines happened to be particularly important in ground-based observations. Analysis of the coronal spectrum has led to suggestions that relative element abundances are not all the same as in the photosphere, although as I have mentioned earlier abundances found from emission lines tend to be less reliable than those found from absorption lines. In particular there appears to be some underabundance of elements with a low first ionisation potential. A long-standing disagreement between the photospheric and coronal iron abundance has, however, recently been shown to be incorrect.

Although the corona could originally only be discovered during a solar eclipse, an invention known as the coronagraph enabled the light from the photosphere to be blocked out by an artificial moon, a disc placed inside the telescope. This meant that the corona could be studied at all times. Apart from the general advantage, this enabled study of the corona outside an eclipse when other indications of magnetic activity suggested that this would be particularly interesting. More recently space observations have enabled some of the coronal line emission to be studied without blocking out the photosphere. Another discovery, which follows from being able to observe the corona over the entire visible face of the Sun in X-rays, is that the temperature of the corona is very non-uniform. The lower temperature regions are known as *coronal holes*. They are particularly prominent near sunspot minimum and near the solar poles and they will be mentioned again in connection with the solar wind.

A major problem is to understand why it is that the Sun possesses a very high temperature outer atmosphere. The density of the solar atmosphere drops off with height very rapidly above the photosphere so that the total mass in the corona is very small. We therefore wish to know why a very small fraction of the mass of the solar atmosphere has a disproportionate share of the energy. For a long time there were only observations of the solar corona, although it was assumed that stars like the Sun would also have one. It is now known that many other late type stars possess coronae and this will be discussed in Chapter 6.

The chromosphere lies between the photosphere and the corona and this was also first observed in solar eclipses. Immediately above the photosphere the kinetic temperature of the matter decreases until it passes through a temperature minimum of 3500 K to 4000 K at a height of a few hundred kilometres. It then rises gradually through the chromosphere and then much more rapidly in the *transition region* until the coronal temperature is reached, as has been shown in fig. 8. It was in the spectrum of the chromosphere that the spectral line of helium was observed before the element was discovered on Earth. I will say more about the chromosphere later. Here I will just mention two other spectral lines which are prominent in the chromosphere. These are the so-called H and K lines of singly ionised calcium. These lines are seen as absorption lines in the spectrum of the solar photosphere and they were given the labels H and K when the strong solar spectral lines were first listed before the relation of spectral lines to particular chemical elements was known. The H and K lines from the chromosphere are emission lines and their strength varies through the sunspot cycle, the lines being stronger at sunspot maximum. H and K emission can be observed from other stars and this is used as evidence of the existence both of stellar chromospheres and of stellar magnetic activity. This will also be discussed in Chapter 6.

The solar wind

It is now known that the Sun is continually losing mass into space. This loss of mass is called the *solar wind*. The existence of intermittent mass outflows from the Sun was first suggested in an attempt to understand *magnetic storms* on the Earth. The properties of the Earth's ionosphere were observed to be modified and radio communication seriously disrupted some time (typically 36 hours) after the observation of some violent activity on the Sun, such as a *solar flare*, to be discussed shortly. What was observed in the Earth's ionosphere could not be caused by electromagnetic radiation from the Sun because that only takes 8 minutes to reach the Earth. It was therefore suggested that the Sun was emitting particles which caused the magnetic storm when they reached the neighbourhood of the Earth.

There were subsequently in the 1950s two separate arguments that the Sun might be emitting ionised matter all the time rather than just intermittently at the time of magnetic storms. The first suggestion arose from observations of comet tails. Comet tails are produced when a comet is close enough to the Sun for the solar radiation to have an important influence on the structure of the comet. To a first approximation the tail of a comet points directly away from the Sun, so that it is behind the comet as it approaches the Sun and in front of it as it recedes. It was originally thought that the production of the tail was all due to radiation pressure. If small particles in the comet absorb radiation from the Sun, they take up not only the energy possessed by the radiation but also the momentum. If they subsequently emit radiation, they will on average emit it equally in all directions and this will carry off no momentum in total. As a result the matter is pushed away from the Sun and produces the *dust tail* of the comet. Observations showed that comet tails were actually divided into two which pointed in slightly different directions (see fig. 15). One tail was the dust tail, which I have just described. The other was shown spectroscopically to be composed of ionised gas and was called the *plasma tail*. It was suggested that this could be understood if the Sun was itself continually

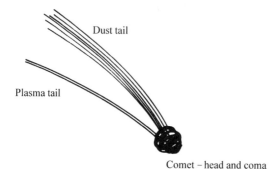

Plasma tail

Comet – head and coma

Fig. 15.   A schematic drawing of a comet showing both a dust tail and a plasma tail.

emitting a stream of plasma. The ionised solar gas would then collide with atoms in the comet and do two things. It would give the particles momentum in the same way that the electromagnetic radiation interacted with the dust particles. There would also be what are called charge exchange reactions in which an electron would be exchanged between an incoming charged particle and a neutral cometary particle leading to the production of a plasma tail, of charged particles moving away from the Sun. Because, as will be discussed in Chapter 4, charged particles move around magnetic field lines, the plasma tail is aligned with the local interplanetary field.

The second suggestion for a continuous solar wind was theoretical. It was observed that the corona was very hot and the question was asked whether a static hot corona could sit at rest at the top of the Sun's atmosphere. I shall discuss this problem in detail in Chapter 5. Here it suffices to say that E N Parker showed that it was impossible for the Sun to support a static high temperature corona. It would exert such a high pressure that it must necessarily expand continuously into the interplanetary medium, leading to a steady mass loss from the Sun. Parker's prediction was made just as it was becoming possible for experiments to be placed on artificial satellites or other space probes and it was very soon discovered that the solar wind does exist.

The solar wind varies in strength through the solar cycle and also erratically at all times. It has an average speed at the Earth of about 400 kms^{-1}. The total mass loss rate is a few times 10^{-14} $M_\odot$ yr^{-1}. This means that, if the solar wind has not been very much stronger in the past, the total loss of mass is only about 10^{-4} $M_\odot$, so that its effect on solar evolution can probably be neglected. The solar wind mass loss is comparable with that due to nuclear reactions in the Sun. It is however very important in studies of the outer solar atmosphere and in the interplanetary medium. In particular, when it reaches the Earth, it confines the Earth's magnetic field in what is known as a *magnetosphere*. It is now believed that it is variations in the strength of the solar wind produced by a surge of magnetic activity on the Sun, which leads to the magnetic storms which I mentioned earlier. This will be discussed in Chapter 7. In terms of the coronal structure which I mentioned earlier, the solar wind flows along the open magnetic field lines which pass through coronal holes, while the highest temperature coronal matter is associated with the closed loops. The changing structure of the corona during the solar cycle leads to the variation in the strength of the solar wind mentioned above. Although Parker's original discussion of the solar wind was of a steady spherically symmetric outflow, the true picture is geometrically much more complicated and variable. Nevertheless, the original qualitative picture survives.

In addition to the solar wind, which is always present but variable, the Sun also loses mass erratically in what are known as *coronal mass ejections*. These eruptive ejections of mass may be comparable in total with the steady solar wind and they appear to be associated with events occurring lower down in the atmosphere, which propagate up into the corona. Some, but not all, accompany solar flares.

Solar flares

A solar flare involves an intense short-lived emission of radiation from a small region of the solar surface. The first recorded observation of a flare was a local brightening of the continuous spectrum of the Sun but most of the emission is observed in spectral lines with Hα, the first line of the hydrogen Balmer series (fig. 16) being the most important in ground based observations. Although the emission from the area of the flare is increased greatly above its undisturbed value, a flare produces a negligible effect on the overall visual or bolometric luminosity of the Sun. A large flare might release 10^{25} J in half an hour. This is very different from the flares on some other low mass stars of spectral type M, which will be discussed in Chapter 6. Their visible luminosity can be dramatically altered, but not their bolometric luminosity, which is principally radiated in the infrared.

Although flares were first discovered through their optical emission, they produce effects throughout most of the electromagnetic spectrum. In particular there are radio and X-ray emissions. The X-rays and ultraviolet radiation provide evidence that locally very high temperatures are produced during a flare outburst. There is of course always X-ray emission from the solar corona but a flare produces a very significant increase where it occurs. The radio waves in contrast indicate that a small fraction of the flare particles are accelerated to high energies. Much of the radiation can be identified as synchrotron radiation, which is produced by electrons which are moving in helical paths around magnetic field lines. All accelerated charged particles radiate and this particular form of radiation is so-named because it is observed in the particle accelerator known as the synchrotron. The flux of high energy particles, cosmic rays, at Earth is also increased as a result of an intense solar flare. This indicates that there may be a stellar origin for many of the cosmic rays of lowest energy. Finally magnetic storms on Earth often follow about 36 hours after flare activity. This is interpreted as being caused by an enhancement in the solar wind which compresses the magnetosphere and increases the magnetic field at the Earth's surface.

Observations of solar flares in all regions of the electromagnetic spectrum from visible to γ-rays provides a wealth of information from which it is possible to unravel the spatial and temporal properties of a flare region. The details are, however, far too complex for

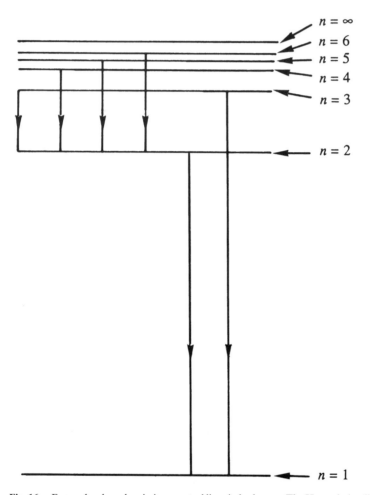

Fig. 16. Energy levels and emission spectral lines in hydrogen. The Hα emission line is produced by electrons moving from the $n = 3$ level to the $n = 2$ level. Transitions to the $n = 1$ level produce the Lyman series in the ultraviolet.

this book and I therefore turn next to the general question as to why flares occur and where the flare energy comes from. The thermal energy in the region of the solar atmosphere in which a large flare occurs is inadequate by several orders of magnitude to provide the flare energy. Thus the energy release in a flare region might be of order 10^2 Jm^{-3}, whereas the thermal energy density in the region of the atmosphere where the flare is situated is only of order 10^{-3} to 10^{-1} Jm^{-3}. The only adequate source of energy is the magnetic energy associated with a strong magnetic field, as a field strength of 1.5×10^{-2} T corresponds to an energy density of 10^2 Jm^{-3}; the energy density of a magnetic fields is $B^2/2\mu_0$, where B is the magnitude of the magnetic induction and μ_0 is the permeability of free space.

It is not enough to say that the magnetic field might possess the energy required to power a flare. It is also necessary to explain why and how the energy is released. It is required that the distribution of magnetic field in the solar atmosphere can change rapidly to one which has a significantly lower energy. The important observation was that flares occur in regions where there is a rapid change in direction of the local magnetic field. In particular there is a neutral line along which one or more components of the magnetic field are zero. In Chapter 4 it will be explained that, near a neutral point in the magnetic field, the timescale for ohmic dissipation of the currents producing the magnetic field becomes very short and a rapid release of energy can occur. The favoured mechanism, which is known as *magnetic reconnection* will be discussed in Chapter 4 and applied to the Sun in Chapter 5.

Solar prominences

Another surface phenomenon of the Sun which was first observed in eclipses is the prominence. Prominences are great areas of luminous material extending outwards from the solar photosphere in the manner shown in fig. 17. The location of the visible prominence changes with time, moving outwards from the solar surface. Although the observation could in principle be one of stationary material with a wave (of temperature variation and hence of luminosity) passing through it, velocity measurements show that

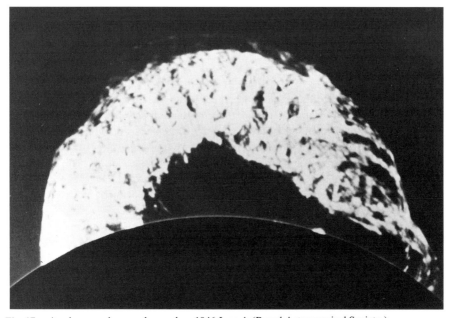

Fig. 17.   A solar prominence observed on 1946 June 4. (Royal Astronomical Society.)

the material is rising. Although the shape of the prominence relative to the whole Sun can only be observed when it is on the edge of the Sun, it is possible to see what happens to prominences as the Sun rotates so that they appear superimposed on the photosphere. Here, contrary to what might have been expected they appear as dark filaments rather than as bright prominences. This indicates that the prominence material is cooler than the photosphere. The largest prominences can have a dimension which is more than a quarter of the radius of the Sun.

Some prominences are short lived eruptive events but others can be quiescent and survive for several rotation periods of the Sun. It is then necessary to understand the equilibrium and stability of the prominences as well as how they are brought into existence in the first instance. It seems clear that the prominence must be threaded along magnetic field lines and that the prominence loop must be embedded in a hotter more diffuse medium. Indeed the upper parts of prominences are often in the region of the hot corona. Simple theoretical ideas suggest that a cool prominence threaded along a magnetic field line which forms a loop convex upwards should not be stable. It is not obvious why the material does not fall down the field lines to the solar surface. However, prominences which are long-lived certainly exist, so that there must be a solution to this problem. As I shall explain in Chapter 6, prominence type activity has been observed in other stars. In the case of the best studied one, AB Doradus, the star is much more rapidly rotating than the Sun and centrifugal forces apparently play a rôle in prominence support and the prominences extend very much further from the stellar surface.

The comment made above that prominences are made of material which is cooler than the photosphere must be qualified. This applies to the quiescent prominences. Eruptive or short-lived prominences can be very much hotter.

Summary of Chapter 2

A brief discussion is given of the properties of stars of all types in preparation for a description of solar properties. Stars can be placed in luminosity classes and spectral types and their properties can be plotted in a Hertzsprung–Russell diagram in which distinct classes of stars appear – main sequence stars, giants, supergiants and white dwarfs. Main sequence stars obey a mass–luminosity relation. Some important groups of stars are not apparent in the HR diagram of nearby stars because there are too few of them or they are too faint to be readily observed; these include pre-main-sequence stars and neutron stars. Stars possess important secondary characteristics which can be deduced by study of stellar spectra. These include rotation and magnetic fields, hot outer atmospheres (coronae) and mass loss. Information about stellar chemical composition can also be obtained. Spectra also disclose the properties of close and interacting binary stars. Because all stars other than the Sun appear as point sources of radiation, direct observations of spatial variations of their surface properties are not possible. Such variations may become apparent as a result of stellar rotation or through eclipse or occultation.

The Sun is a main sequence star of spectral type G. Its mass, radius and luminosity are known accurately and it is observed that its luminosity varies very slightly through the sunspot cycle, with maximum luminosity coinciding with maximum sunspot number. The age of the solar system is known rather accurately and it is generally believed that the Sun and the solar system formed together, so that the Sun has essentially the same age. The chemical composition of the outer layers of the Sun can also be determined, although there are some elements which do not have spectral lines excited in the solar photosphere from which reliable abundances can be obtained. A particularly important example is helium, even though it was first detected through observations of the solar chromosphere.

Although the average properties of the Sun are essentially time independent (or more accurately they are believed to be varying on a timescale of about 10^9 yr), there are small scale variations in both space and time. Energy transport just below the solar surface is principally by convection and this produces a pattern of rising and falling elements, the granulation, in the photosphere. There are also small amplitude vertical oscillations with a characteristic period of five minutes. In addition there are all of the phenomena associated with surface magnetic activity; sunspots, solar flares and prominences and the hot corona and the solar wind.

3 The quiet Sun and stellar evolution

The equations of stellar structure

A major concern of astronomers is to understand the observed properties of stars and how these properties have changed in the past and will change in the future. This has been discussed in some detail in the companion book *The Stars: Their Structure and Evolution* but I shall summarise the principles here.

In most phases of stellar evolution, the structure of a star can be determined by the solution of four first order ordinary differential equations. In the case that energy transport in the star is by radiation, the four equations take the form

$$\frac{dP}{dr} = -\frac{GM\rho}{r^2}, \tag{3.1}$$

$$\frac{dM}{dr} = 4\pi r^2 \rho, \tag{3.2}$$

$$\frac{dL}{dr} = 4\pi r^2 \rho \epsilon, \tag{3.3}$$

$$\frac{dT}{dr} = -\frac{3\kappa L\rho}{16\pi acr^2 T^3}. \tag{3.4}$$

In these equations r is the distance from the stellar centre, P, ρ, T are the pressure, density and temperature at radius r, M the mass contained within radius r, L the energy carried by radiation across radius r, ϵ the nuclear energy release $kg^{-1}s^{-1}$, κ the opacity, which measures the resistance of the material to energy transport by radiation, and G, a, c, the constant of gravitation, radiation density constant and velocity of light. It is implicit in equations (3.1) to (3.4) that the star being considered is spherical and that its properties are changing very slowly with time. The validity of these two assumptions will be discussed below, when it will be explained that they are very good for the Sun at the present stage in its evolution.

The three quantities P, ϵ and κ depend only on the chemical composition and physical state of the stellar material. I can therefore write schematically

$$P = P(\rho, T, \text{composition}), \tag{3.5}$$
$$\epsilon = \epsilon(\rho, T, \text{composition}), \tag{3.6}$$
$$\kappa = \kappa(\rho, T, \text{composition}). \tag{3.7}$$

The determination of these expressions for P, ϵ and κ is pure physics and requires no knowledge of astronomy. However, because physical conditions inside stars are very different from those on Earth, their study has generally been left to astronomers. There are uncertainties in the values of these quantities which will be mentioned later.

Convection

Equation (3.4) must be modified and supplemented if there is any significant energy transport by convection and this is the case in at least part of most stars. In (3.4), L must be replaced by L_{rad} the energy carried by radiation and a further equation is required for L_{conv} the energy carried by convection. Thus

$$\frac{dT}{dr} = -\frac{3\kappa L_{\text{rad}}\rho}{16\pi a c r^2 T^3}, \tag{3.4'}$$
$$L = L_{\text{rad}} + L_{\text{conv}}, \tag{3.8}$$
$$L_{\text{conv}} = ?. \tag{3.9}$$

I have written (3.9) in the form shown to indicate that there is not at present a good theory of convection, although, as explained in *The Stars*, the precise form of the theory proves to be unimportant in deep stellar interiors. The uncertainty in the theory may be important in stellar atmospheres and I shall return to that point later in this chapter. Energy transport by conduction is unimportant in most stages of stellar evolution except for very low mass stars, but when it is important it can be included in equation (3.4) with an appropriate modification to the form of the opacity.

Convection is an instability. If an element of material is displaced upwards and is then lighter than its surroundings it will continue to rise. The criterion for the onset of convection is obtained by assuming that the rising element moves sufficiently slowly that it is in pressure balance with its surroundings but that at the same time its motion is adiabatic so that no heat is exchanged between the element and the surroundings. Convection then occurs if

$$\frac{P}{\rho}\frac{d\rho}{dP} < \left(\frac{P}{\rho}\frac{d\rho}{dP}\right)_{\text{ad}}, \tag{3.10}$$

where the second term is the value for adiabatic motion. If the stellar material is an ideal classical gas with constant ratio of specific heats γ, the inequality becomes

$$\frac{P}{\rho}\frac{d\rho}{dP} < \frac{1}{\gamma}. \tag{3.11}$$

If, moreover, the chemical composition and state of ionisation of the material are uniform with depth, the criterion takes the form,

$$\frac{P}{T}\frac{dT}{dP} > \frac{\gamma - 1}{\gamma}. \tag{3.12}$$

In this form it can be seen that convection occurs if the temperature gradient required to carry all of the energy by radiation exceeds a critical value. All of this is discussed in *The Stars*. In the solar convection zone, the ionisation of hydrogen and helium, the two most abundant elements, is variable with depth and, as a result, the two simpler forms of the criterion for the onset of convection are invalid but inequality (3.10) is always applicable.

Dynamical and thermal timescales

Although equations (3.1) to (3.4), with the possible modifications just mentioned, suffice for a study of most stars, there are rapid stages of stellar evolution in which additional terms involving time derivatives must be included in equations (3.1) and (3.3). In obtaining (3.1), it has been assumed that the star is in *hydrostatic equilibrium*. If the forces on any element of stellar material are not in balance, it is accelerated and a term $\rho \partial^2 r / \partial t^2$ must be inserted in equation (3.1). In addition, equation (3.3) does not allow for the possibility that the energy released in or passing through an element of material may cause it to expand or heat up. Terms allowing for these two possibilities are sometimes required in (3.3). They are particularly important in the earliest stages of the life of a star when the interior has not become hot enough for nuclear reactions to occur. Then it is gravitational energy released by contraction, which provides the energy that the star radiates.

The acceleration of stellar matter is only important, if the star's properties are changing in a time not much greater than the *dynamical timescale*, which is approximately

$$t_D = (r_s^3 / GM_s)^{\frac{1}{2}}, \tag{3.13}$$

where r_s and M_s are the total radius and mass of the star. This is the time required for a star to collapse to a small radius if the pressure is unimportant. For the Sun, $t_{D\odot} \simeq \frac{1}{2}$ hr. The heating and expansion terms are only important if the stellar properties change on the *thermal or gravitational timescale*.

$$t_{Th} = GM_s^2 / L_s r_s, \tag{3.14}$$

where L_s is the total luminosity of the star. In such a time a star radiates an amount of energy roughly equivalent to its thermal energy content. For the Sun $t_{Th\odot} \simeq 10^7$ yr. It is believed that the average properties of the Sun are currently not changing much in 10^9 yr. In consequence it should be possible to understand gross solar properties using the equations already described. I shall however explain that the Sun does exhibit some changes on each of the dynamical and thermal timescales and that these are important in understanding the fine details of the structure of the Sun and its relation to other stars.

Effects of rotation and magnetic fields

There is a further respect in which the equations so far described may not be adequate. It has been assumed that stars are spherical. This will not be the case if a star is rotating very rapidly, contains a strong magnetic field or is distorted by interaction with a close binary companion. It does not appear that any of these factors is important in determining the overall properties of the Sun at the present time. The surface rotation of the Sun is unimportant in the sense that the ratio of the centrifugal force to gravitational force acting on an element of matter at the Sun's equator satisfies the inequality

$$\Omega^2 r_s^3 / G M_s \ll 1, \tag{3.15}$$

where Ω is the observed solar angular velocity. A comparison of (3.13) and (3.15) shows that the condition is equivalently that the solar rotation period should greatly exceed the dynamical timescale. That is obviously easily satisfied as the rotation period is about 28d and the dynamical timescale about $\frac{1}{2}$ hr. The observed surface magnetic field of the Sun is very important in determining the detailed surface properties, as I shall discuss in later chapters of this book. The value of the internal field of the Sun, which is deduced in the most straightforward manner as an extension of the observed surface field, possesses a total energy which is very much less than the gravitational and thermal energies of the Sun. Such a magnetic field is incapable of changing the overall properties of the Sun.

There have been suggestions that the deep interior of the Sun might be rotating much more rapidly than the outside or that there might be a very strong magnetic field in the solar core. In the latter case most of the field lines of this strong magnetic field must close inside the Sun so that the flux is not visible coming through the solar surface. Both of these suggestions were made in an attempt to understand two observational solar properties. One of the observations has been shown to be incorrect as a result of later careful studies. Rotating self-gravitating bodies are flattened at the poles because centrifugal force resists gravity in the equatorial plane but not along the axis of rotation. If inequality (3.15) is satisfied, the flattening is very small, as is observed to be the case for both the Sun and the Earth. A measurement of the solar flattening suggested that, although it was small, it was larger than could be explained by the observed surface rotation. The additional flattening might be explained either if the solar interior rotated very much more rapidly than the surface or if the Sun contained a very strong internal magnetic field. When the solar flattening, or oblateness as it is called, was remeasured and found to be consistent with the surface angular velocity, this placed constraints on the speed of internal rotation and on the strength of any internal magnetic field.

The second observation, which will be discussed later in this chapter, is that of neutrinos emitted in the nuclear reactions converting hydrogen into helium, which are believed to provide the luminosity of the Sun. The number of neutrinos detected has always been lower than the number expected according to theory. The simplest explanation, when the discrepancy was discovered, was that the central temperature of the Sun was lower than theory predicted. This could be true if either a strong magnetic field or rapid internal rotation helped the gas pressure to resist gravity. A lower central gas pressure would then be required and this would imply a lower central temperature.

I shall explain later in this chapter that neither of these is believed to provide a true resolution of the solar neutrino problem. One reason for this belief is that a further method, which I shall discuss, has become available for studying the internal structure of the Sun. The surface of the Sun exhibits low level oscillations which can be interpreted as the normal modes of free oscillation of the Sun; these are the five-minute oscillations mentioned in Chapter 2. The spectrum of normal modes depends on the internal solar structure. Some oscillations are determined by the properties of the surface regions of the Sun alone, while others are determined by the structure of the deep interior. In addition both rotation and a magnetic field can produce mode splitting similar to the Zeeman effect in spectral lines mentioned in Chapter 2. Study of solar oscillations indicates that the solar interior is not rotating very rapidly and that there is not a very strong internal magnetic field. Another argument against a strong field is related to the stability of the field itself and this will be discussed in the next chapter.

Although rotation and magnetic fields are unimportant in determining the overall structure of the present Sun, they are crucial in determining its detailed surface properties as I shall explain in Chapter 5. They also are crucial in the interaction of the Sun with the Earth and with other magnetised bodies in the solar system. I shall discuss solar-terrestrial relations in Chapter 7. It also seems likely that the Sun rotated much more rapidly and possibly that it had a much stronger dipole component of its magnetic field earlier in its life history. This will be discussed in Chapter 6 when I consider the properties of younger solar-type stars and in Chapter 8 when I consider star formation and the origin of the solar system.

Calculation of solar structure

I now return to the study of the present Sun with what seems the safe assumption that I can describe its overall properties by the simplest set of equations stated earlier in the chapter, apart from the need to allow for energy transport by convection in the outer layers. To obtain the properties of any star through a solution of these equations, I require more items of information. What star I am studying requires a value for the total mass of the star and a specification of the chemical composition of that material. In addition the equations can only be solved if boundary conditions for these equations are provided. In discussing a solution of the equations, it is simplest to consider them in a form in which the mass, M, is used as the independent variable, because if the total mass, M_s is specified, the appropriate range of values for M (0, M_s) is known. Equations (3.1) to (3.4) can be rewritten

$$\frac{\mathrm{d}P}{\mathrm{d}M} = -\frac{GM}{4\pi r^4}, \tag{3.16}$$

$$\frac{\mathrm{d}r}{\mathrm{d}M} = \frac{1}{4\pi r^2 \rho}, \tag{3.17}$$

$$\frac{\mathrm{d}L}{\mathrm{d}M} = \epsilon, \tag{3.18}$$

$$\frac{\mathrm{d}T}{\mathrm{d}M} = -\frac{3\kappa L}{64\pi^2 a c r^4 T^3}. \tag{3.19}$$

As explained earlier equation (3.19) needs modification when convection is occurring.

Boundary conditions

These four ordinary differential equations need four boundary conditions. Two of these apply at the centre of the star, $M = 0$, and two at $M = M_s$. The two at the centre are unambiguous and self-evident. Thus

$$r = L = 0 \text{ at } M = 0. \tag{3.20}$$

The star can neither have a hole nor a point source of luminosity at its centre. The surface boundary conditions are more complicated. Even though the Sun appears to have a sharp edge, when studied visually, it actually has an extensive outer atmosphere as I have explained in Chapter 2. The same is presumably true of other stars. The first and simplest approach to a stellar surface boundary condition is to argue that, as the surface density and temperature of the Sun (and other stars) are much less than the average density and temperature inside the Sun, it should be possible to obtain a reasonable solution of the equations by taking

$$\rho = T = 0 \text{ at } M = M_s. \tag{3.21}$$

If a value for the actual surface temperature of the Sun is required, the effective temperature can be found from the equation

$$L_s = \pi a c r_s^2 T_e^4. \tag{3.22}$$

The simple surface boundary conditions (3.21) are often adequate for a study of stellar structure, but this is not true for low mass stars with extensive outer convection zones. In this case better surface boundary conditions are required. The form of the next approximation has been described in *The Stars*. The conditions are

$$L = \pi a c r^2 T^4, \quad P\kappa = 2g/3 \text{ at } M = M_s. \tag{3.23}$$

where g is the surface gravity GM/r^2. The boundary conditions (3.21) or (3.23) are meant to be adequate for a determination of the gross properties of a star such as its radius, luminosity and effective temperature. The detailed determination of more detailed observed atmospheric properties such as spectrum, surface motions, mass loss, surface activity is then the subject of a separate investigation of the surface layers in which these gross properties are assumed known.

A calculation of the present properties of the Sun proceeds as follows. The first model calculated is known as a zero age main sequence model (ZAMS). A ZAMS star is one which has just settled on to the main sequence with nuclear reactions converting hydrogen into helium providing its surface energy release. Because gravitational energy release, which powered pre-main-sequence evolution, is now unimportant, the structure of the star is determined by solving the differential equations in which no time derivatives appear. Provided the mass and chemical composition of the star are specified, the equations can be solved without any knowledge of the previous properties of the star. This is very important because even today both star formation and pre-main-sequence evolution are imperfectly understood.

Solar evolution

Stellar evolution immediately following the ZAMS can be studied by supplementing the equations which I have already listed with equations describing how the chemical composition of the stellar material changes as a result of nuclear reactions. It is usually assumed, for reasons which I shall describe shortly, that the chemical composition of a ZAMS star is homogeneous. Nuclear reactions are highly temperature-dependent and only occur in the central regions. As a result hydrogen is gradually converted into helium in the centre. If there is no convection in the stellar core, which is believed to be the case in the Sun, the change in composition is frozen in the material in which it occurs. If I define

$$X = \text{fraction by mass of material in form of hydrogen (}^1\text{H)}, \tag{3.24}$$

$$Y = \text{fraction by mass of material in form of helium (}^4\text{He)}, \tag{3.25}$$

I can then write

$$\left(\frac{\partial Y}{\partial t}\right)_M = -\left(\frac{\partial X}{\partial t}\right)_M = f(\rho, T, \text{composition}). \tag{3.26}$$

In fact it is a little more complicated than this because I should really consider the build up and destruction of intermediate isotopes such as deuterium (^{2}H) and ^{3}He. The principle is not changed. The function on the right hand side of (3.26) is obviously closely related to ϵ. If convection occurs in the central regions of a star, the products of the nuclear reactions will be mixed through the convective region. Convection currents are usually so rapid that the convective core may be considered to have uniform composition. It is then not difficult to find an equation to replace (3.26).

In order to obtain a theoretical model for the present structure of the Sun, it is now necessary to specify the mass and chemical composition of the Sun when it reached the

ZAMS and the time that has elapsed since then (its age). I consider each of these in turn. The present mass of the Sun is known quite accurately ($M_\odot = 1.99 \times 10^{30}$ kg) as is its luminosity ($L_\odot = 3.83 \times 10^{26}$W). The mass of the Sun is currently changing for two reasons. The nuclear reactions in the solar centre are converting mass into energy to provide the luminosity of the Sun. There is thus a mass loss rate which I can write

$$\dot{M}_\epsilon = -L_\odot/c^2. \tag{3.27}$$

In fact the true mass loss rate is a few per cent higher than this because neutrinos emitted in the conversion of hydrogen into helium carry off some of the released energy. There is also the mass loss in the form of the solar wind. If the observed mass loss in the solar wind and the deduced mass loss from nuclear reactions are added together, it appears that the mass of the Sun should not change significantly in less than about 10^{13} yr. As the current age of the Sun is thought to be more like 4.6×10^9 yr and, in any case, the age of the Universe is believed to be less than 2×10^{10} yr, there should be no error in assuming that the ZAMS Sun had the same mass as the present Sun. There are much larger uncertainties in solar structure than that.

Solar chemical composition

In considering the composition of the ZAMS Sun, there are three separate questions. Was the composition of the ZAMS Sun uniform? Have there been any changes in solar surface composition since then? What is the present composition of the solar surface? The final question has been discussed in the previous chapter where I have explained that there are indeed some uncertainties in the present composition of the solar surface. It is not a simple matter to convert the observed strength of spectral lines into element abundances, even if all the relevant atomic parameters are known. There continue to be inaccurately known atomic parameters. Some elements do not have spectral lines in the visible region of the solar system. Fortunately similarities in the chemical composition of the Sun and other objects in the solar system (Earth, meteorites) enables an estimate to be made of the solar abundance of some other elements. This will be discussed further in Chapter 8. The major outstanding problem is helium which cannot be observed in the solar photosphere and which must largely have escaped from the Earth and the meteorites. As a result the solar helium abundance cannot even today be regarded as well known from direct observation.

It is generally assumed both that the ZAMS Sun was of homogeneous composition and that the present surface composition is that original composition. I will describe briefly the reasons for those beliefs and also what might cause them to be wrong. The Sun formed out of an interstellar gas cloud which probably contained a much greater mass than the Sun. Many stars are observed to form in giant molecular clouds with masses of order $10^5 \, M_\odot$. The interstellar gas in the Galaxy has a chemical composition which is slowly changing as a result of the expulsion from stars of material, which has been enriched in heavy elements by nuclear reactions in the stars. Theoretical studies of the manner in which mass expelled by stars mixes with the ambient interstellar medium suggest that the interstellar gas and the star-forming clouds that form out of it should have broadly homogeneous chemical composition. If one solar mass of material is plucked out of such a cloud to form a star, it will probably have a rather smooth composition.

If the material which was to form the Sun had uniform composition, is there any way in which the composition could have changed other than through nuclear reactions? Here there are two general possibilities. The first is that the present surface composition may be different from that of the interior because the surface has been modified by accreted contaminating matter. The second is that some separation of chemical elements has occurred in the Sun. Consider the first possibility. The Sun travels in an orbit around the centre of the Galaxy, each orbit taking about 2.4×10^8 yr. In its orbit the Sun might pass from time to time through a dense interstellar cloud and pick up some cloud material, which would change its surface composition. The total mass of such material would be very small compared to $M_\odot$. It seems very unlikely that the present surface composition is that of a contaminating cloud with a very different composition from the solar interior for at least three reasons. The outer regions of the Sun are stirred by convection currents so that a small mass of accreted material should rapidly be diluted in the much larger mass of the convection zone. In addition the solar wind, which would have to be cut off to allow accretion to occur, is continually removing the outer layers and, so to speak, is keeping the Sun clean. Finally the very close similarity between the chemical composition of the Sun and other bodies in the solar system would be a coincidence if such dirtying of the solar surface had occurred.

The other possibility is that some elements have settled relative to others in the gravitational field of the Sun. The argument about the similarity in composition of the Sun and the solar system is less strong here because, as I have mentioned, not all elements can be observed in the Sun. Gravitational settling is apparently occurring in some stars more massive than the Sun, particularly in those known as Am stars, which are observed to have peculiar surface chemical compositions. However, these are stars without an outer convection zone in which there is not believed to be any mixing process which can interfere with the settling. It is generally believed that the outer convective zone is kept well-mixed and that this prevents the observed surface composition of the Sun from deviating from the original composition. The position in the deep solar interior, where there is no convection is perhaps less clear, but for the moment I will assume that an initially homogeneous solar composition has only been changed by nuclear reactions.

The solar age

If I assume that the mass and initial chemical composition of the Sun are known, I can solve the equations of stellar structure and calculate the course of solar evolution. To compare theoretical properties with those of the real Sun, I must stop the integration at the present age of the Sun. What is that age? As I shall explain in more detail in Chapter 8, it is believed to be about 4.6×10^9 yr. This value is obtained from the assumption that the entire solar system formed at the same time. Studies of the abundance of heavy radioactive elements and of their decay products in meteorites and in lunar and terrestrial rocks enable the time since the rocks solidified to be calculated. The oldest samples which have been studied, including many of the meteorites, are about 4.55×10^9 yr old. If the solar system was still forming at the time the Sun reached the main sequence, the Sun could well be about 4.6×10^9 yr old. Earlier theories of the formation of the solar system had the planets forming as the result of the interaction of the Sun with another star. If such theories were correct the Sun could be significantly older than 4.6×10^9 yr. Such theories currently do not appear very plausible, and I shall assume that the solar age is about 4.6×10^9 yr. Some authors favour an age more like 4.5×10^9 yr believing that meteorites could have solidified before the Sun settled on the main sequence as a hydrogen burning star, but this only introduces a two per cent uncertainty in the solar age.

Standard solar models

If the integration of the equations of stellar structure is stopped after 4.6×10^9 yr, we should like it to predict that, at that time the radius of the Sun, $r_\odot$, is 6.96×10^8 m and the luminosity, $L_\odot$, is 3.83×10^{26} W. It is highly unlikely that this will be found to be true. The reason is that there are uncertainties in the input to the solar model. These uncertainties are in:

(i)   The value of the solar age;
(ii) The solar initial composition;
(iii) Values of P, κ and ϵ, even for specified composition;
(iv) Energy transport by convection for which there is not a good theory.

The next question is whether, within the generally accepted uncertainties of (i) to (iv), a solar model can be found which has the observed solar surface properties. There is no serious problem in satisfying this requirement and the solution is not unique. Different scientists have tabulated their own *standard solar models* which have the required observable properties and which incorporate what they regard as the best set of values for the quantities (i) to (iv). Different standard solar models do not agree in fine detail but they are generally similar. Authors tend to revise and refine their standard models as improvements are made in expressions for such quantities as P and κ. Important recent changes have arisen as a result of modifications in the accepted values of κ at temperatures below about 10^6 K.

Solar convection zone

A particular problem arises from the lack of a good theory of convection. Immediately below the visible solar surface there is a region in which a significant amount of energy transport is by convection. This zone occupies almost a third of the solar radius but only contains about two per cent of the mass. To a close approximation it can therefore be regarded as a region in which the mass and luminosity have their surface values M_s and L_s. A major reason for the occurrence of convection is that hydrogen and helium are neutral at the solar surface but that they are ionised just below the surface. In these ionisation zones the ratio of the specific heat at constant pressure to the specific heat at constant volume is much lower than the value 5/3, which is appropriate either to a neutral gas or to a fully ionised gas, and this encourages convection.

As I have explained in *The Stars* there is not at present a good theory of energy transport by convection. The theory which is usually used, the *mixing length theory*, contains a free parameter. This mixing length is usually written

$$l = \alpha H_p = \alpha |P/(dP/dr)|, \tag{3.28}$$

where H_p is the *pressure scale height*. It is supposed that α is order of unity. The structure of the solar convection zone depends on the value of α. In particular it influences the depth of the bottom of the convection zone as well as the total radius of the Sun. Other information about the probable depth of the convection zone comes from the study of solar oscillations which I shall describe shortly. Using this and other information it is possible to determine a best value of α to use in a description of the solar convection zone and hence in a complete model of the Sun. The value obtained in recent standard solar models is about 2. This is higher than the value accepted a few years ago.

It has to be recognised that this may be the best value of α assuming that the mixing length theory provides an adequate description of convection in the Sun but that the properties of convection described by the theory may not closely resemble the actual pattern of solar convection. Even if the concept of a mixing length is valid, its relation to the pressure scale height may vary both with position in a star at any given time and with time as the properties of the star change. Once the observed properties of the Sun have been used to calibrate the mixing length theory, it is natural in the first instance to use the same value of α in other stars with outer convection zones, but this may not, in fact, be appropriate.

Convective motions in stars do more than transport energy. They also keep the convection zone well stirred so that it has an essentially uniform chemical composition. This is not very important in the solar convection zone because the outer layers of the Sun are in any case expected to have uniform composition. I have, however, already mentioned that convection probably prevents any attempt of heavy chemical elements to settle in the Sun's gravitational field or any stratification produced by radiation pressure. Convection also carries around angular momentum and, in a highly conducting fluid, magnetic flux and this is crucially important in the outer convection zones of the Sun and other stars, as I shall discuss in Chapters 5 and 6.

So far everything can be regarded as extremely satisfactory. There is a good general agreement between theory and observation and this can be expected to improve as better knowledge is obtained relating to the physical quantities which enter the equations and to the age and composition of the Sun. Any standard solar model not only provides values for quantities at the solar surface but also gives a complete run of these quantities through the solar interior. At the time such models were first obtained, it did not seem possible that any of the theoretical predictions concerning the internal properties of the Sun could be tested by observation. It is now realised that there are at least two techniques which can in principle give knowledge about the deep solar interior. The first involves the detection of neutrinos from nuclear reactions and has given rise for more than 25 years to what is known as the *solar neutrino problem*. The second is the study of solar oscillations which can probe the solar interior through helioseismology in much the same way that earthquake waves enable study of the Earth's interior. It is with these two topics that much of the rest of this chapter is concerned.

Energy generation in the Sun

There are two reaction chains which can convert hydrogen into helium in stellar interiors. The first is known as the Proton–Proton Chain (PP) and the second is known as the Carbon–Nitrogen Cycle (CN). The PP chain involves the direct build up of protons into helium nuclei whereas the CN cycle uses carbon and nitrogen nuclei as catalysts. The PP chain dominates in stars with low central temperature and the CN cycle is more important at high temperatures. It is believed that almost all of the energy released in the main sequence Sun comes from the PP chain.

The PP chain has three branches. The main branch, through which most of the reactions go is shown in expression (3.29).

$$p(p, e^+ + \nu_e)d(p, \gamma)^3He(^3He, 2p)^4He.^* \tag{3.29}$$

In this expression p stands for a proton, d for a deuteron (nucleus of heavy hydrogen), e^+ a positron, ν_e an electron neutrino and γ a photon (or gamma ray). In this chain three protons combine successively to form a helium-3 nucleus and then two of these ^{3}He nuclei form ^{4}He with the release of two surplus protons. The net effect of the chain, as of any reaction converting hydrogen to helium, is

$$4p \rightarrow {}^4He + 2e^+ + 2\nu_e. \tag{3.30}$$

The positrons and neutrinos must be produced because the helium-4 nucleus contains two protons and two neutrons and the conversion of a proton to a neutron produces a positron and a neutrino. The positrons rapidly annihilate with electrons and produce

---

* The abbreviated notation used here can be explained as follows. If a reaction occurs in which, for example, two particles A and B combine to form two other particles C and D, I write this A(B,C)D. The initially present particles are written to the left of the comma and the final particles to the right.

γ-rays, which are in turn rapidly absorbed, but the neutrinos can escape essentially freely from the Sun.

The other branches to the chain arise because when ^{3}He is produced it can react with ^{4}He which is already present in the star. The possible chains are shown in (3.31).

$$^3\text{He}(^4\text{He}, \gamma)^7\text{Be}(e^-, \nu_e)^7\text{Li}(p, {}^4\text{He})^4\text{He}$$
$$(p, \gamma)^8\text{B}(, e^+ + \nu_e)^8\text{Be} \rightarrow 2^4\text{He}. \tag{3.31}$$

In these chains a choice arises when ^{7}Be can either capture an electron (e$^-$) or react with a proton. At the end of the third chain ^{8}Be is unstable and breaks up to 2^4He. The second chain is less probable than the first but, in turn, it is much more probable than the third. There is one further slight variant of the first chain, a three particle reaction between two protons and an electron which produces a deuteron and a neutrino

$$p + e^- + p \rightarrow d + \nu_e. \tag{3.32}$$

A final very rare reaction is

$$^3\text{He} + p \rightarrow {}^4\text{He} + e^+ + \nu_e. \tag{3.33}$$

The reaction of two deuterons to produce ^{4}He can be ignored because the deuterons are destroyed too rapidly by collisions with protons.

As far as the overall structure and evolution of a star is concerned the neutrinos emitted in the above reactions are not very important. They carry away a few per cent of the energy released in the conversion of hydrogen into helium and, as they escape essentially freely from the star, this energy is not available to supply its surface luminosity. As a result there is a corresponding small reduction in the time for which hydrogen burning supplies the star's luminosity. The neutrinos emitted in reactions (3.29) and (3.31)–(3.33) have different energies. What is important in what follows is that the rare neutrinos emitted by the decay of ^{8}B are much more energetic than the others apart from the even rarer neutrinos from the ^{3}He + p reaction.

The CN cycle is not believed to be important in the Sun, although it is the main source of energy in massive main sequence stars and it should become important in the Sun when the Sun becomes a red giant. The principal reactions in the CN cycle are shown in expression (3.34)

$$^{12}\text{C}(p, \gamma)^{13}\text{N}(, e^+ + \nu_e)^{13}\text{C}(p, \gamma)^{14}\text{N}(p, \gamma)^{15}\text{O}(, e^+ + \nu_e)^{15}\text{N}(p, {}^4\text{He})^{12}\text{C}. \tag{3.34}$$

The key point in this reaction cycle is that almost always the capture of a proton by ^{15}N produces ^{4}He and ^{12}C. The ^{12}C nucleus is now available to start a new cycle and the net effect is again shown in (3.30). The final reaction very rarely proceeds as follows

$$^{15}\text{N}(p, \gamma)^{16}\text{O}(p, \gamma)^{17}\text{F}(, e^+ + \nu_e)^{17}\text{O}(p, {}^4\text{He})^{14}\text{N}, \tag{3.35}$$

which produces ^{4}He and re-enters the main chain at ^{14}N rather than at ^{12}C. Neutrinos released by decay of ^{13}N, ^{15}O and ^{17}F are predicted to make a finite but very small contribution to the neutrinos emitted by the Sun. Additional reactions to (3.34) and

(3.35), which involve more massive nuclei, only occur in more massive stars or at later stages of stellar evolution.

Solar neutrinos

The neutrinos emitted in the first reaction of the PP chain have an energy which can have any value between zero and 0.42 MeV, whereas the maximum energy from decay of ^{8}B is about 15 MeV. The energy spectrum and flux of solar neutrinos which should be received at Earth according to a standard solar model is shown in fig. 18. The data shown in this figure are relatively insensitive to small changes in the standard model. It can be seen that the neutrinos from ^{7}Be and pe$^-$p have discrete energies whereas all of the others have a continuous distribution. All neutrinos interact very weakly with matter but the probability of absorption does increase with energy. Thus the rare ^{8}B neutrinos are more likely to be absorbed than the lower energy neutrinos. The probability of any process is measured in terms of an absorption cross-section. This is the effective area offered by a target particle to a beam of incident particles. The actual absorption is then the same as would be the case if all particles hitting the area were absorbed and all others

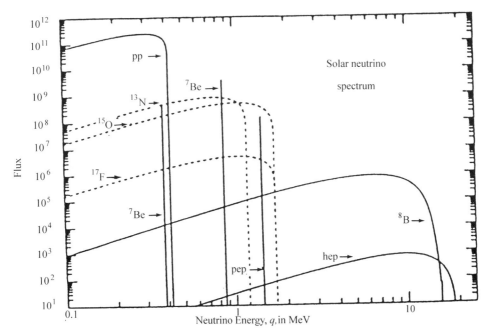

Fig. 18. The solar neutrino spectrum for the standard solar model of Bahcall and Ulrich (*Reviews of Modern Physics*, **60**, 297, 1988) showing that the majority of high energy neutrinos are released by decay of ^{8}B. The fluxes from continuum sources are in units of number cm^{-2} s^{-1} MeV^{-1} at the position of the Earth and the line fluxes in number cm^{-2} s^{-2} (1 MeV $\approx$ 1.6 $\times$ 10^{-13} J).

escaped freely. The actual cross-section for the interaction of neutrinos with matter depends both on the energy of the neutrinos and on the composition of the target, but an order of magnitude estimate for the interaction of a typical solar neutrino with a target is

$$\sigma \simeq 10^{-50} \, \text{m}^2. \tag{3.36}$$

This compares with cross-sections more like $10^{-20} \, \text{m}^2$ and $10^{-30} \, \text{m}^2$ in atomic and nuclear physics. The distance between collisions, if the target particles have a number density $n \, \text{m}^{-3}$, is

$$l = 1/n\sigma \simeq 10^{21} \, \text{m}, \tag{3.37}$$

where I have taken $n \simeq 10^{29}$, a reasonable value for a solid target. This distance is large even by astronomical standards, showing that neutrinos are indeed very weakly interacting.

Despite the very large value of the mean free path in equation (3.37), there is a possibility of detecting solar neutrinos because there are so many of them. To within less than a factor 2 there should be $10^{15} \, \text{m}^{-2} \, \text{s}^{-1}$ reaching the Earth. With the cross-section and number density given above, the number of detections $\text{m}^{-3} \, \text{s}^{-1}$ would be

$$r \simeq 10^{-6} \, \text{m}^{-3} \, \text{s}^{-1} \tag{3.38}$$

or about one detection per month per cubic metre of detector. This is a very crude estimate which depends on there being a practicable detector which does present a cross-section of order the value given in (3.36) to typical solar neutrinos. That was not the case in the first solar neutrino experiment which I shall now describe. The crude calculation does, however, indicate that an attempt to detect solar neutrinos is not doomed to failure.

Solar neutrino experiments

The original experiment involved the capture of neutrinos by the less abundant isotope of chlorine, ^{37}Cl; this accounts for about a quarter of natural chlorine. It had been shown that this isotope had a higher than average cross-section for the absorption of neutrinos and that it was abundant and easy to handle. The reaction is

$$^{37}\text{Cl} + \nu_e \rightarrow \, ^{37}\text{Ar} + e^-. \tag{3.39}$$

The ^{37}Ar produced is unstable and subsequently decays

$$^{37}\text{Ar} \rightarrow \, ^{37}\text{Cl} + e^+ + \nu_e, \tag{3.40}$$

and it is the decay of ^{37}Ar which is detected. The disadvantage of ^{37}Cl as a target is that ^{37}Ar is sufficiently more massive than ^{37}Cl (that it must be more massive is shown by reaction (3.39)) that only neutrinos with energy greater than 0.8 MeV can initiate the reaction. This rules out the most numerous low energy neutrinos and the majority of the captures must involve the rare ^8B neutrinos, which also have a higher cross-section than the typical value stated above. This was regarded as an advantage rather than a disadvantage when the experiment was mounted in the 1960s, for a reason which I shall explain shortly.

The ^{37}Cl is in a tank containing 10^5 gallons of C_2Cl_4 (perchloroethylene) in the Homestake Gold Mine in Lead, South Dakota. The experiment has to be placed deep underground to avoid contaminating reactions produced by cosmic rays. Because argon is an inert gas, it is possible to extract it from the tank and to observe its decay elsewhere. A full account of the experiment can be found, for example, in the book *Neutrino Astrophysics* by J N Bahcall referred to in the Suggestions for Further Reading at the end of this book.

When the ^{37}Cl solar neutrino experiment was first suggested by R Davis (following an earlier outline proposal by B Pontecorvo) it was believed that it would provide a very good discrimination between different theoretical models for the structure of the Sun. As explained earlier, within the present physical uncertainties, there is a range of models which provide agreement with the observed surface properties of the Sun but they differ in their value for the solar central temperature, $T_{c\odot}$. All of the models must provide the observed luminosity, $L_\odot$, and in consequence they predict the same flux of neutrinos from the initial reaction of the PP chain, because the number of these reactions is directly related to the solar luminosity. The number of reactions involving ^{8}B is in contrast very sensitive to $T_{c\odot}$, being approximately proportional to $T_{c\odot}{}^{18}$. The Davis experiment was expected to detect neutrinos according to the then standard model of the Sun. The precise number detected should then provide a better estimate of the solar central temperature and refine the standard solar model. This was not what happened.

Comparison of experimental and theoretical neutrino fluxes

Neutrino detections are measured in the solar neutrino unit defined by

$$1 \text{SNU} \equiv 10^{-36} \text{interactions sec}^{-1} \text{target atom}^{-1}. \tag{3.41}$$

The experiment contains about 2×10^{30} ^{37}Cl atoms so that a detection rate of 1 SNU would imply one detection about every 5×10^5 s or about 6 days. The original theoretical value of the detection rate was 30 ± 15 SNU obtained by Bahcall in 1966. In fact, the initial result, once a clear detection was achieved, was that far fewer neutrinos were detected than expected, with the implication that the central temperature of the Sun was lower than given in any of the models which gave the correct surface properties.

This did not initially cause great alarm because it was generally accepted that there might be greater uncertainty in the application of physical laws to solar structure than was currently estimated. Theoretical astronomers were quickly able to produce new models of the Sun with a lower predicted neutrino flux. However, this initial decrease was not sustained and there has continued to be a serious discrepancy between theory and experiment. In summary, it does not seem that allowing the expressions for P, κ and

ϵ, the value of the mixing length and the solar chemical composition to vary within their currently accepted uncertainties can produce a solar model which has both the correct solar surface properties and a low neutrino flux in the chlorine experiment. Two later theoretical models predicted the following count rates:

$$\text{Bahcall and Pinsonneault (1992) } 8.0 \pm 3.0 \, \text{SNU},$$
$$\text{Turck-Chièze and Lopes (1993) } 6.4 \pm 1.4 \, \text{SNU}. \tag{3.42}$$

Although these values look quite different, the solar models are quite similar. The high dependence of the ^{8}B neutrino flux on $T_{c\odot}$ exaggerates the difference. In contrast the current experimental result is

$$2.28 \pm 0.23 \, \text{SNU}. \tag{3.43}$$

The difference between theory and experiment appears to be well outside their uncertainties.

Non-standard solar models

There have been further experimental developments but before I describe these I will say a few brief words about the attempts that were made for more than 20 years to reconcile theory and experiment. There have been many proposed non-standard models. Some of these were not a priori implausible but some of them looked like an act of desperation almost as soon as they were proposed. The suggestions can be grouped as follows:

(i)   There is an additional force resisting gravity in the solar interior which reduces the central pressure and temperature – rapid internal rotation, strong internal magnetic field.

(ii)   The Sun's main energy is not nuclear – the Sun contains a central black hole or neutron star, with gravitational energy release from accretion providing much of the radiated energy.

(iii) The Sun is not completely stable. The instability mixes chemical composition and causes the present nuclear reaction rate and neutrino flux to be lower than the average.

(iv)   The surface chemical composition of the Sun is not typical of the interior composition.

(v)   There is a mechanism of energy transport in the solar interior which competes with radiation and reduces the temperature gradient – waves, weakly interacting massive particles (WIMPs).

All of these suggestions other than (ii) have had periods of popularity. If (i), (iv) or (v) were correct the neutrinos emitted by the initial PP reaction should arrive with the flux predicted by the standard solar model. This is not true for (ii) or (iii) but for (iii) to be valid the release of nuclear energy must vary with a period less than the solar thermal timescale ($\sim 10^7$ yr) so that the surface luminosity is not characteristic of the present rate of energy release. The suggestion that the Sun might contain a small quantity of weakly interacting massive particles, with a mean free path for interaction comparable with but less than the solar radius, had some attractions. There is much hidden mass in the Universe and probably in the Galaxy, which could be such weakly interacting particles, and the Sun could capture some of them at a steady rate as it travels though them. With their large mean free path they could carry energy effectively through the solar interior and reduce the temperature gradient and $T_{c\odot}$.

The concern with trying to resolve the solar neutrino problem was not simply one of trying to understand the properties of the Sun. There was first the general problem of whether there was something radically wrong with the whole understanding of stellar structure; this was not regarded as very plausible because of the successes of the subject, but it had to be a background worry. The second concern was whether the Sun had properties specific to itself or to its situation in the Galaxy, which meant that it could not be regarded as typical of all stars of $1\,M_\odot$ and with solar age and composition. This might be the case if the Sun had an atypically large internal rotation or magnetic field or if the solution was caused by WIMPs. The density of WIMPs and hence the capture rate by stars would depend on position in the Galaxy and would vary from galaxy to galaxy.

All of the non-standard solar models are currently out of favour and the general belief is that it is the neutrino which is the culprit. One reason for this is the additional information about the solar interior provided by solar oscillations, which I shall discuss later in the chapter. Before I discuss the properties of the neutrino, I will describe additional solar neutrino experiments which have supplemented the results of the ^{37}Cl experiment. The most important results are provided by two experiments which use gallium as a target. The advantage of gallium is that it is sensitive to some of the low energy neutrinos released by the PP reaction. The problem with gallium is that unlike chlorine it is a rare element and the amount of gallium required to mount a successful experiment was soon realized to be greater than the amount of pure gallium produced worldwide in a year. Whereas perchloroethylene is a readily available cleaning fluid, gallium is expensive and can only be obtained in suitably large amounts by international collaboration.

The gallium neutrino experiments

The reaction which is used is

$$\nu_e +\,^{71}\mathrm{Ga} \to \mathrm{e}^- +\,^{71}\mathrm{Ge}. \tag{3.44}$$

Neutrinos with an energy $\geqslant 0.2332\,\mathrm{MeV}$ can initiate this reaction and this includes many of the PP neutrinos. ^{71}Ge decays through electron capture, the reverse reaction to (3.44). In atomic form it captures one of its own electrons with a half-life of 11.43 d. Two experiments have been mounted which have been given the names SAGE and GALLEX. The SAGE experiment is a Soviet/American experiment (or rather was before the break-up of the USSR) and it uses 60 tons of metallic gallium. The GALLEX experiment is a largely European experiment located underground in Italy. It uses 30 tons of gallium in a GaCl$_3$-HCl solution. In each case similar difficult experimental techniques must be used to measure the amount of germanium produced. Bahcall's text *Neutrino Astrophysics* is once again a good source of detailed information.

More than half of the neutrinos which can be detected by gallium come from the PP reaction with the second most important contribution being made by neutrinos from ^{7}Be. This indicates that the gallium experiments complement the chlorine experiment. The predicted capture rates for gallium for two recent standard solar models are:

$$\text{Bahcall and Pinsonneault (1992) } 132 \pm 7\,\text{SNU},$$
$$\text{Turck-Chièze and Lopes (1993) } 123 \pm 7\,\text{SNU}. \tag{3.45}$$

These two results are in agreement to within their uncertainties and it is unlikely that any other standard solar model would give a figure substantially different from either of these.

At the time of writing the gallium experiments, like the chlorine experiments, are providing results which do not agree with the theoretical predictions, although not by such a large factor. The latest results which I have are

$$\text{GALLEX } 79 \pm 12\,\text{SNU},$$
$$\text{SAGE } 58 \pm 24\,\text{SNU}. \tag{3.46}$$

It does not seem possible that all these results can be accommodated by assuming that the central temperature of the Sun is indeed smaller than current calculations predict for a reason that is not at present understood. The gallium experiments have not been running long enough for it to be clear that the results given in (3.46) will be the final ones.

Kamiokande

The Kamiokande experiment was not originally designed to study solar neutrinos. The experiment involves a large tank of pure water sited underground and its aim was to study the possible decay of the proton, which is predicted by some proposed theories of elementary particles. Although the half-life, if the proton is unstable, is known to be substantially in excess of 10^{30} yr, it is possible to study the decay if a sufficient mass of protons can be studied. The Kamiokande experiment came into prominence as a neutrino detector when it picked up a number of neutrinos from the explosion of the supernova SN 1987A in the Large Magellanic Cloud. The device has subsequently been used in its enhanced form, Kamiokande II, to study solar neutrinos. This contains 0.68 kilotons of water. Kamiokande can only detect ^{8}B neutrinos as the threshold energy is 7.5 MeV. The detection arises by neutrino scattering on electrons

$$\nu_e + e^- \rightarrow \nu'_e + e^{-\prime}. \tag{3.47}$$

In this case it is possible to verify that the neutrinos are coming from the Sun because the angular distribution of the scattered electrons gives information on the direction of the incident neutrinos. The result of the Kamiokande experiment is also inconsistent with theoretical predictions with a discrepancy by a factor of about 2.

Further experiments

At the time of writing some additional solar neutrino experiments are planned. The Kamiokande detector is being upgraded and as a result its neutrino threshold will be reduced to 5 MeV. Another experiment SNO will scatter neutrinos on deuterium as well as on electrons. The neutrino scattering on deuterium has two possible outcomes

$$\nu_e + {}^2H \rightarrow p + p + e^-, \tag{3.48}$$

and

$$\nu + {}^2H \rightarrow n + p + \nu. \tag{3.49}$$

I have deliberately not put a suffix e on the neutrino in (3.49). This reaction can be induced by the muon neutrino, ν_μ, and the tauon neutrino, ν_τ, as well as by ν_e. Only ν_e is emitted by the solar nuclear reactions but one possible explanation of the results of current experiments, which I shall discuss below, is that ν_e can change into either ν_μ or ν_τ between the centre of the Sun and the Earth. As current experiments are not sensitive to ν_μ or ν_τ, this might account for the deficit of detected neutrinos. If neutrino flavour changing does occur, the number of neutrinos which can produce reaction (3.48) will be reduced relative to those which can induce (3.49) and this should be apparent in the results. A further experiment BOREXINO is intended to detect, amongst other things, the line emission of neutrinos from ${}^7Be$ at 0.86 MeV.

Possible explanation of results

There is now a widely held hope that the resolution of the solar neutrino problem may involve developments in particle physics rather than in astrophysics. In the current standard model of particle interactions there are three absolutely conserved quantities called electron, muon and tauon lepton numbers and correspondingly three types of neutrino all of which have zero mass and are completely stable. Particle physicists, however, believe that there are likely to be modifications to the standard model. Most of the suggested modifications involve the neutrinos having small masses, which are consistent with the present experimentally established upper limits to their masses, and also involve the possibility that a neutrino of one type can transform into a neutrino of another type.

The currently favoured idea is that a mechanism of neutrino oscillations known as the Mikheyev–Smirnov–Wolfenstein effect causes the flux of electron neutrinos at Earth to be less than that expected from the number emitted from the centre of the Sun. There are some parameters in the MSW effect, which will only be known if the true form of the particle interactions is established, but there is a range of values of the parameters which is capable of explaining the Chlorine, Gallium and Kamiokande results. It remains to be demonstrated that the MSW effect does exist and that its parameters fall in the right range. Indirect evidence may be provided by the SNO experiment mentioned above. If this proves to be the resolution of the solar neutrino problem, it has to be recognised that the solar neutrino experiments will have told us less about the solar interior than about the laws of particle physics. It would, however, give us greater confidence in applying the predictions of theories of stellar evolution to other stars of masses close to that of the Sun.

There is one further highly speculative idea about the influence of neutrino properties on the measured solar neutrino flux. It has been suggested, but not very convincingly, that the flux is variable in the solar cycle. This might be explained if the neutrino possessed a magnetic moment and if interactions with the solar magnetic field produced mixing of different types of neutrino. This proposal has several problems associated with it. It is first of all rather unlikely that the neutrino does have a large enough magnetic moment to produce a significant effect. In addition any variation through the solar cycle would depend on changes in the magnetic field throughout the solar interior and it is not at all obvious what changes in the internal field are associated with variations in solar surface activity.

Solar oscillations

As I mentioned in Chapter 2, the five-minute solar oscillations are identified as the normal modes of oscillation of the Sun. A full description of solar oscillations is complex and will not be attempted but a brief account of the theoretical ideas is given in Appendix 4. A first approximation in a study of the oscillations is the assumption that the Sun is strictly spherical. This should provide a spectrum of oscillation frequencies which will be modified slightly by perturbing factors such as the solar rotation and the solar magnetic field. A second approximation is the assumption that the oscillatory motions are adiabatic which means that no heat flows into or out of any mass element during the oscillation. This should be a good approximation given that the oscillation period is very much smaller than any relevant thermal timescale. A third approximation is a neglect of the change in the gravitational field of the Sun during the oscillation. This cannot be strictly accurate for radial oscillations in which all the matter at any solar radius moves inward or outward in phase, but it should be a good approximation for non-spherical modes, particularly those with a short wavelength in the horizontal direction.

It is explained in Appendix 4 that there are two different types of oscillation which are possible in the Sun. In one of them the most important restoring force is pressure and in the other it is gravitation. The first set of modes is known as p-modes and the p-modes have frequencies between about an hour and two minutes; these include the five minute oscillations that were originally observed. The other modes are known as g-modes and

Fig. 19. Motions on the solar surface due to p-modes. The white parts correspond to outward motions and the black parts to inward motions (from Turck-Chièze *et al. Physics Reports,* **230**, 59, 1993).

they have much longer periods than the p-modes. The g-modes are trapped in the solar interior beneath the convection zone. Their amplitude is damped exponentially in the convection zone so that their surface effects are very slight and so far no detections have been made of g-modes. I will therefore concentrate my discussion on what has been learnt by study of the p-modes.

Any oscillation is labelled by three quantum numbers n, l and m. The first quantum number corresponds to the number of points in the radial direction at which the amplitude of the oscillation vanishes. The quantum numbers l and m (where $-l \leqslant m \leqslant +l$) determine the angular behaviour of the oscillation over the surface of the Sun. The inward or outward motion of points on the surface is related to the value of the real part of the function $P_l^m(\cos\theta)\exp(im\varphi)$, where θ, φ are spherical polar coordinates and where P_l^m is known as an associated Legendre function and its analytical form and numerical values are well known. Thus if l and m are low, there is a relatively small number of patches on the solar surface with different directions of the radial velocity; if l and m are large there is a very large number of such patches. This is illustrated in fig. 19. A model is said to have a high degree if l is large and conversely if l is small. Most of the observable p-modes have periods between 2 and 10 minutes, with 5 minutes as a characteristic value.

As is also explained in Appendix 4, the p-modes are also trapped in a region of the Sun with the top somewhere near to the solar surface and with the position of the bottom depending on the properties of the speed of sound in the solar interior and on the value of l. For small values of l the waves can penetrate almost to the centre of the Sun, whereas for high values of l the modes are trapped close to the surface. The oscillation frequency of any mode depends on the internal properties of the Sun in the region in which the mode can propagate. This means that in principle the observation of the oscillations can provide information about the manner in which quantities vary in the solar interior.

Having said that two challenging problems remain. The first one is that of observing the solar oscillations and of identifying the different modes. The second is that of inverting the observed data in order to deduce the internal properties of the Sun. Consider first the observations. The most obvious observational technique is to measure the Doppler shift of the wavelength of solar spectral lines which is produced by motion towards or away from the observer. This is an extremely delicate task. Observed spectral lines are always broadened for several reasons including particularly the thermal motions of the absorbing or emitting particles and any mass motions such as convection or turbulence. In the case of solar spectral lines, the spread in line-of-sight velocity of absorbing particles is up to $10\,\mathrm{km s}^{-1}$. In contrast the velocities associated with individual p-modes are more like 0.1 to $0.2\,\mathrm{m s}^{-1}$. It is clear that a measurement of such small additional velocities requires very precise and stable instrumentation. There are other observational techniques which involve measuring brightness variations on the solar surface which accompany the oscillations or apparent changes in the solar diameter. The former has been done successfully but the latter will probably only produce useful results if the observations are made from spacecraft.

If the different oscillations are to be resolved two things are necessary. There must be a time-resolution of the observations which is adequately shorter than the oscillation period and there must be a good spatial coverage of the solar surface. It is then possible to extract from the observations oscillation frequencies corresponding to modes with different spherical harmonic degree l. I shall not describe the details of how this is done but will simply discuss the results. It is possible to observe thousands of such modes and the results of one investigation are shown in fig. 20. There are many modes corresponding to each value of l; these have different values of n. If the Sun were non-rotating there would be no variation of oscillation frequency with m. Because the solar rotation period is so much longer than the oscillation period, this is still approximately true. There is an m splitting but that is very much smaller than the n splitting.

An observation of the m splitting is important in attempting to study the internal rotation of the Sun. If the Sun were rotating like a solid body the rotational splitting would be quite simple. However, it is known from observations of the solar surface that even that does not have uniform rotation. If the angular velocity varies with depth in the Sun, the mode splitting must depend particularly on the rotation of the regions of the

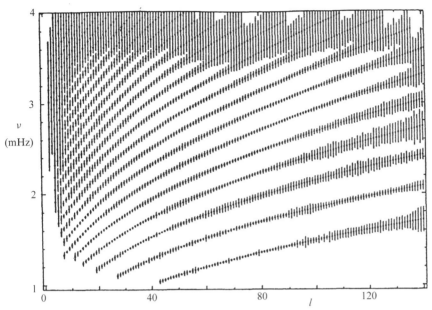

Fig. 20. Solar p-mode frequencies from Libbrecht & Woodard, *Nature*, **345**, 779, 1990. The vertical lines overestimate the probable errors. The distinct sequences of frequencies correspond to different values of *n*. The lowest frequency sequence has $n = 1$.

Sun in which the mode is concentrated. As have explained low degree modes penetrate almost to the solar centre whereas high degree modes are restricted to the surface layer. This means that a variation in splitting between modes of different *l* can provide information about how the solar rotation varies with depth. Note that if the Sun contained a well-ordered strong internal magnetic field, such as a dipole, this would also produce an *m*-splitting. The problem would however become complicated, if both rotational and magnetic splitting were present, unless their axes of symmetry coincided. All of the evidence is that, contrary to some suggestions that have been made in the past, there is no strong field in the solar interior.

I now leave the observational techniques to discuss the way in which the results are used to study the internal structure of the Sun. Clearly two approaches are possible in principle. Assuming that there is a well-defined standard solar model such as has been used in predicting the solar neutrino flux, all of its relevant oscillation frequencies can be calculated from the equations discussed in Appendix 4. Although this approach is straightforward, the chance that the calculated frequencies will agree with the observed frequencies is slight. A continuation of this approach would involve changing the parameters of the solar model slightly and recalculating the frequencies. This would make sense if there were just a few well-defined parameters which defined a solar model. In contrast, because of uncertainties in such quantities as chemical composition, opacity and the convection theory, there is in effect a continuum of such parameters. This leads to use of the second technique which is an inversion of the observed data to try to determine the run of physical quantities in the solar interior. This is called helioseismology because it is akin to the way in which the properties of earthquake waves are used to probe the interior of the Earth.

Because the methods are complex I shall only describe them in broad outline. In any case the techniques used by various groups differ in detail. The general idea is, however, quite clear. As explained in Appendix 4 the observed oscillation frequencies all depend on how quantities such as density and temperature vary throughout the solar interior, although some frequencies are really only sensitive to the values of these quantities close to the surface. The variations in temperature and density in turn depend on the input data such as the chemical composition of the Sun and the fine details of such things as the opacity and the convective energy transport. If I assume first that the Sun is spherically symmetric and ignore the rotational splitting of the oscillations, I can assign values of the relevant physical quantities such as ρ and T at a number of radial positions and calculate the oscillation frequencies to reasonable accuracy. Alternatively I can regard the values of ρ and T etc. as the unknowns and use the observed oscillation frequencies to obtain values of ρ and T. The total number of quantities which I can determine is then equal to the number of observed oscillation frequencies for which the values of l and n can be regarded as well known. This means that, if more frequencies can be identified, a better model of the internal structure of the Sun can be obtained. How well a technique like this works depends on the oscillation frequencies providing independent information about the solar interior, which is why one needs to study both low l and high l so that knowledge can be obtained about the central regions as well as the surface regions.

Once the frequencies have been inverted to obtain a model for the internal structure of the Sun, it is then necessary to compare the model with those obtained from solutions of the equations of stellar structure. Here use can be made of experience in solving those equations because it is known which regions of the Sun are most sensitive to such things as uncertainties in the chemical composition and the laws of opacity and energy generation. Thus any distinct differences between the model obtained by seismic inversion and the preferred solar model used in discussions of the the solar neutrino problem may occur in a region where a first guess may readily be made for the cause of the discrepancy. Does it perhaps arise from the wrong assumed value of the solar helium content or from the use of the wrong value of the mixing length in the theory of convection? If a useful hint is obtained, then new standard models can be calculated in the hope that they will lie closer to the inverted model.

There is also great interest in studying the rotational splitting of the different l modes as this can provide information about the internal angular velocity of the Sun. As I have explained in Appendix 4, the rotational splitting would be quite simple if the Sun rotated like a solid body. As it does not, the different rotational splitting of the various modes enables information to be obtained about the angular velocity in the regions in which they are concentrated. It is immediately clear that inverting the data to obtain the angular velocity is more complicated and unreliable than inverting the data to obtain the values of ρ and T. The reason is that we observe that the surface angular velocity is a function of latitude, which means that we must expect the internal angular velocity to be a function of both r and θ. This means that very many more values of the angular velocity, Ω, must be determined than of ρ and T in order to constrain its variation adequately. In fact studies of the internal rotation of the Sun have concentrated on trying to answer two types of question. The first is how does Ω vary with both r and θ in the outer solar convection zone because this is very important in trying to understand the relation between convection, rotation and magnetic fields and in particular the dynamo maintenance of the solar field. The second is whether the central regions of the Sun rotate distinctly more rapidly than the outer layers. I shall return to the topic of the evolution with time of the angular velocity of the Sun and other stars in Chapters 6 and 8.

When the observations of solar oscillations were inverted to determine the solar internal structure, it was found that they predicted a deeper outer convection zone than was found in existing standard solar models. This stimulated much detailed theoretical study of the physics of the outer Sun. In recent years it has become clear that the value of the opacity in the outer regions of stars had been underestimated. If the opacity is higher, so is the temperature gradient needed to carry energy by radiation and this makes convection more likely. The discrepancy between theoretical models and helioseismological results has now been much reduced. Helioseismology also provides an estimate of the solar central temperature. This suggested that it could not be low enough for the WIMP explanation of the solar neutrino discrepancy even before the gallium neutrino experiment indicated that a simple lowering of the central temperature would be unlikely to solve the problem.

The simplest interpretation of the rotational splitting of solar oscillations is that the observed surface differential rotation of the Sun persists throughout the outer convection zone but that there is a rapid change to uniform rotation below the convection zone boundary. It is in this region of rapid velocity shear that the solar dynamo, to be discussed in Chapters 4 and 5, is believed to be most active. Early measurements suggested that the angular velocity rose towards the centre of the Sun but that is now doubtful. In any case a rapid enough rotation to influence the surface shape of the Sun on the result of the neutrino experiment is ruled out.

There should be a considerable increase in the knowledge obtained from solar oscillations in the next few years. There is a large cooperative study by ground-based telescopes known as the GONG network. In addition a satellite SOHO was launched in late 1995.

Before I leave the subject of helioseismology, I must mention the hope that some of these techniques will be successfully used on other nearby stars. It is clear that spatially resolved information cannot be obtained if the star appears as a point source and information must be obtained by integrating over the whole disk of the star. Some groups have made such observations of the Sun which is thus equivalent to observing the Sun as a star. These observations are most sensitive to the $l = 0$, radial oscillations of the Sun although some contribution is also made by low l modes. This means that study of the Sun in this manner is useful for providing information about the structure of the deep interior. This is in fact probably the best way of making such studies.

Summary of Chapter 3

A theoretical study of the structure of the Sun, or any other star, requires the solution of four differential equations, which are concerned with the balance of forces in the star, the release of energy and the transport of energy through the star. These equations must be supplemented by expressions, in terms of the density, temperature and chemical composition at any point in the star, for the pressure, the energy release from nuclear reactions and the opacity, which is inversely related to the energy transport by radiation. There is also need for an expression for the energy transport by convection. The differential equations have their simplest form if a star is spherical and if its properties are slowly varying with time. It can easily be demonstrated that these assumptions are valid to a good approximation for the Sun.

The properties of a star, which has reached the stage where internal nuclear reactions are supplying its radiation, depend on the star's mass, on its chemical composition and on its age. The age here is defined to be the time since nuclear reactions converting hydrogen into helium started and the star became a main sequence star. In the case of the Sun, we have a very good value for its mass. Only its surface chemical composition can be studied and this is known reasonably well. There are arguments that the observed surface composition was the composition of the entire Sun before nuclear reactions started to change the central composition but there is no way of verifying this directly. The age of the Sun is believed to be essentially that of the whole solar system and this is believed to be known to an accuracy of one to two per cent.

A theoretical model of the present Sun is obtained by first studying the zero age main sequence Sun and by then following its evolution to the present age. The model must then give correct values for the current luminosity, radius and effective temperature of the Sun. Given that there are uncertainties in the input physics, it proves not too difficult to produce a solar model with these properties. Such a model is known as a *standard solar model*. A standard solar model provides predictions for the internal properties of the Sun as well as its observed surface properties. There are now known to be two ways in which a study can be made of the solar interior.

In the nuclear reactions converting hydrogen into helium neutrinos are released. They are so weakly interacting that they should almost all escape freely from the Sun. It is possible to attempt to detect solar neutrinos by terrestrial experiments. So far the number detected has always been smaller than the number predicted by any standard solar model. This is a matter of concern because, if it is not possible to understand solar structure and evolution, it is difficult to expect to understand the structure of other stars. Although the neutrino problem has not yet been resolved, the currently most favoured idea is that the discrepancy has nothing to do with astronomy. It seems possible that all of the neutrinos which leave the centre of the Sun do not survive in their original form until they reach the Earth. Instead it is possible that they may change into another type of neutrino, which cannot be detected by the current experiments.

The solar interior can also be probed because its surface exhibits low level oscillations which can be interpreted as normal modes of oscillation of the Sun. The oscillation frequencies are determined by the internal structure of the Sun and it is possible to try to invert the observed spectrum of oscillation frequencies to determine the solar structure. This is known as helioseismology. Helioseismology has already provided useful information about such things as the depth of the solar convection zone and the internal rotation of the Sun and it appears to support standard solar models rather than a very different internal structure.

4   Magnetic fields and fluid motions*

Basic equations

The surface activity in the Sun is strongly influenced by, if not controlled by, the magnetic field in the solar outer layers. In order to discuss the activity and that in other stars it is necessary to have some knowledge of the properties of magnetic fields in general and of their rôle in the Universe. Electromagnetic fields have properties which are governed by four equations known as Maxwell's equations. These can be written

$$\operatorname{curl} \mathbf{H} = \mathbf{j} + \frac{\partial \mathbf{D}}{\partial t}, \tag{4.1}$$

$$\operatorname{curl} \mathbf{E} = -\frac{\partial \mathbf{B}}{\partial t}, \tag{4.2}$$

$$\operatorname{div} \mathbf{B} = 0, \tag{4.3}$$

$$\operatorname{div} \mathbf{D} = \rho_{\mathrm{E}}. \tag{4.4}$$

In these equations $\mathbf{H}$, $\mathbf{B}$, $\mathbf{D}$, $\mathbf{E}$, $\mathbf{j}$ and ρ_{E} are respectively the magnetic field and magnetic induction, electric displacement and electric field, electric current density and electric charge density.

These equations must be supplemented by five further equations. Those relating $\mathbf{B}$ and $\mathbf{H}$ and $\mathbf{D}$ and $\mathbf{E}$ depend, in general, on the presence or absence of permanent magnets and dielectrics but in most gaseous media in the Universe they are adequately approximated by

$$\mathbf{B} = \mu_0 \mathbf{H}, \quad \mathbf{D} = \epsilon_0 \mathbf{E}, \tag{4.5}$$

where μ_0 and ϵ_0 are the permeability and permittivity of free space. A third equation relates the electric current density to the fields producing it. The simplest form of this generalised Ohm's law, which is frequently valid, is

$$\mathbf{j} = \sigma \left[\mathbf{E} + \mathbf{v} \times \mathbf{B} \right], \tag{4.6}$$

---

* A knowledge of vector calculus is required for a full understanding of this chapter. The basic results will be summarised in a manner which is independent of their derivation.

where σ is the electrical conductivity and $\mathbf{v}$ is the bulk velocity of the matter. In equation (4.6) $\mathbf{E} + \mathbf{v} \times \mathbf{B}$ is the electric field felt by a body moving with velocity $\mathbf{v}$ as required by the special theory of relativity. The final equations depend on the microscopic make up of the matter. If it consists of just electrons and one type of ion, the equations are

$$\mathbf{j} = n_i Z_i e \mathbf{v}_i - n_e e \mathbf{v}_e, \tag{4.7}$$
$$\rho_E = n_i Z_i e - n_e e, \tag{4.8}$$

where n_i, $\mathbf{v}_i$ and n_e, $\mathbf{v}_e$ are the number density and velocity of the ions and electrons respectively and $Z_i e$ and $-e$ are the charges on the ion and the electron. In general there will be more than one species of ion and also neutral particles. The presence of fluid and particle velocities in equations (4.6) to (4.8) means that there is a close link between the electromagnetism and dynamics of the system.

Long-lived magnetic fields

In astrophysics it is common to treat magnetic fields as essentially permanent components of a system, whereas electric fields are regarded as transient. There are good reasons for this. Equation (4.3) is a statement that there are no magnetic charges or magnetic monopoles as they are usually known. It is a common experience that division of a permanent magnet into two does not separate north and south poles. What is produced is two smaller magnets each with two poles. The third Maxwell equation enshrines this as a law. There are currently suggestions from particle physics that this might not be totally correct. There may be some magnetic monopoles in the Universe but, even if there are, they are so few that equation (4.3) is still essentially valid. Electric fields can be produced by separating positive and negative charges through equation (4.4) but the attraction between these charges is so strong that the charge separation is usually cancelled out very quickly. There is no corresponding effect involving magnetic fields. Electric fields can be produced by time varying magnetic fields through equation (4.2). Such fields are only significant if there are rapid changes in the magnetic field. Magnetic fields produced by the displacement current, $\partial \mathbf{D}/\partial t$, are usually insignificant in astrophysical problems because electric fields are unimportant, but magnetic fields which are slowly varying both temporally and spatially can be produced by a conduction current $\mathbf{j}$ if the electrical conductivity is high enough. There is no corresponding magnetic current producing electric fields because of the absence of magnetic monopoles. Henceforth I shall omit the term in $\partial \mathbf{D}/\partial t$ in equation (4.1).

I consider next the behaviour of magnetic fields in stationary media. In this case I can combine the equations

$$\operatorname{curl} \mathbf{H} = \mathbf{j}, \quad \operatorname{curl} \mathbf{E} = -\partial \mathbf{B}/\partial t, \quad \mathbf{B} = \mu_0 \mathbf{H}, \quad \mathbf{j} = \sigma \mathbf{E} \tag{4.9}$$

to obtain the single vector differential equation for $\mathbf{B}$

$$\frac{\partial \mathbf{B}}{\partial t} + \frac{1}{\mu_0 \sigma} \, \text{curl curl } \mathbf{B} = 0. \tag{4.10}$$

The vector identity curl curl $\mathbf{B} \equiv \text{grad div } \mathbf{B} - \nabla^2 \mathbf{B}$, together with div $\mathbf{B} = 0$, enables equation (4.10) to be written

$$\frac{\partial \mathbf{B}}{\partial t} = \frac{1}{\mu_0 \sigma} \nabla^2 \mathbf{B}. \tag{4.11}$$

Thus, for example, in cartesian coordinates

$$\frac{\partial \mathbf{B}_x}{\partial t} = \frac{1}{\mu_0 \sigma} \left[\frac{\partial^2 \mathbf{B}_x}{\partial x^2} + \frac{\partial^2 \mathbf{B}_x}{\partial y^2} + \frac{\partial^2 \mathbf{B}_x}{\partial z^2} \right]. \tag{4.12}$$

Equation (4.11) describes the manner in which a magnetic field changes as a result of finite electrical conductivity. In fact, solution of the equations shows that magnetic fields decay together with the currents producing them.

If the currents vary significantly in a distance L, their approximate decay time is

$$\tau_D = \mu_0 \sigma L^2. \tag{4.13}$$

It can be seen very simply from equation (4.12) that, if at time $t = 0$ there exists a sinusoidal field

$$B_x = B_0 \exp(iky), \tag{4.14}$$

the solution at later times is

$$B_x = B_0 \exp(iky) \exp(-k^2 t / \mu_0 \sigma). \tag{4.15}$$

The wavelength λ of the spatial variation of the field is $2\pi k$ and the original field decays by a factor e in the time $\mu_0 \sigma \lambda^2 / 4\pi^2$. When it is recognised that the true distance over which the field varies significantly is $\lambda/4$ rather than λ, this time is very close to that given in (4.13).

The solar magnetic field

In normal laboratory experiments currents decay rapidly not because the electrical conductivity is low but because the thickness of the material in which they are flowing is small. In astrophysical situations L is very large and in a star the electrical conductivity of a fully ionised gas, or plasma, is also high, comparable with that of copper. If L is the radius of a main sequence star it is found that for stars like the Sun or more massive the lifetime of a magnetic field could exceed the main sequence lifetime. Such a field that has been present in a star possibly since its formation is called a *fossil field*. The same is not true for the Earth whose field is believed to be produced by currents in a liquid conducting core. Here expression (4.13) predicts that any fossil field would long since have decayed. The terrestrial field must therefore be continuously regenerated. The process that does this is called a dynamo mechanism and I shall discuss it shortly. The electrical conductivity of an ionised gas is proportional to $T^{3/2}$ which means that the characteristic time for the decay of currents in the outer layers of the Sun is very much less than the solar lifetime, whereas the decay time near the centre exceeds the lifetime. If the field in the solar interior were a fossil field extending throughout the Sun, the field in the outer layers would now be current free such as the field produced by a dipole. This is not what is observed. The surface field is very complex, as I have already described in Chapter 2, and it must also be regenerated by a dynamo action. It is in any case clear that the discussion which I have given so far cannot apply in detail to the outer layers because of the convective motions which occur there. Thus it is inappropriate to omit the term $\mathbf{v} \times \mathbf{B}$ in equation (4.6).

Before I discuss other properties of magnetic fields in stars, I will say a few more words about a possible strong magnetic field which might exist in the interior of the Sun which is beneath the convection zone. As I have stated in Chapter 3 this region contains most of the mass of the Sun and occupies about 75 per cent of the solar radius. If there were a strong magnetic field trapped in this region when the star reached the main sequence, it could survive serious decay until the present time. It is not clear whether there could be such a fossil field. Some ideas about pre-main-sequence stellar evolution suggest that the Sun might have been fully convective at one stage and that the convective motions could have tangled up the magnetic field and enhanced its decay rate once the motions ceased. This position is not clear and it is useful to have other ideas about the possible existence and survival of a strong interior field. As I have already explained in Chapter 3, helioseismology argues against a strong field. I now give a second argument.

Magnetic buoyancy

Another argument is related to the stability of such a field. It is in fact possible to show that most very strong magnetic fields trapped within a star are unstable. The full discussion is very complex and I shall only describe one particular instability, which will be important in what follows. This is magnetic buoyancy which is illustrated in fig. 21. If magnetic flux is concentrated as shown and embedded in a region with which it is in thermal balance, it will only be in equilibrium with its surroundings if the total pressure inside the flux tube is equal to the pressure outside. A magnetic field in a conducting fluid exerts a force per unit volume which is

$$\mathbf{F}_{\text{mag}} = \mathbf{j} \times \mathbf{B} = (\text{curl}\,\mathbf{B} \times \mathbf{B})/\mu_0. \tag{4.16}$$

This force can be rewritten

$$\mathbf{F}_{\text{mag}} = -\text{grad}(B^2/2\mu_0) + \mathbf{B}.\nabla\mathbf{B}/\mu_0. \tag{4.17}$$

The first term in (4.17) is the gradient of an isotropic pressure and the second term represents a tension along the lines of magnetic induction. The magnetic pressure must

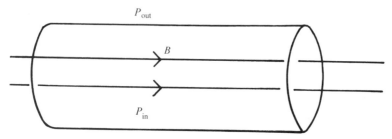

Fig. 21. Magnetic buoyancy. A tube of magnetic flux with magnetic induction B has an internal pressure P_{in} which must be smaller than the external pressure P_{out}. Because temperature is continuous, the density in the flux tube is lower than that of its surroundings and it rises.

be added to the gas pressure in considering the continuity of pressure across the interface shown in fig. 21. Thus

$$P_{out} = P_{in} + B^2/2\mu_0. \tag{4.18}$$

The gas pressures can be written $\mathscr{R}\rho T/\mu$ where $\mathscr{R}$ is the gas constant, ρ the density and μ the mean molecular weight of the stellar material. With $T_{in} = T_{out}$ we must have

$$\rho_{in} < \rho_{out}. \tag{4.19}$$

A tube of magnetic flux is lighter than its surroundings and will therefore rise in just the same way that material less dense than water floats on water. Although I have demonstrated this buoyancy effect for an isolated tube of magnetic flux it can be generalised to other situations in which a strong magnetic field is buried. This and other instabilities which I shall not describe seems likely to rule out the continued existence of a strong internal field even if there was originally one. I shall return to the idea of magnetic buoyancy later.

Magnetic flux freezing

An important property of magnetic fields in conducting fluids is that to a good approximation the fluid is tied to the magnetic field. This means either that the fluid motions drag the magnetic field lines around or that the magnetic force given by equation (4.16) is so strong that it constrains the motion of the fluid. Both of these possibilities arise in different regions of the outer atmosphere of the Sun. In some cases fluid is constrained to move along magnetic field lines, whereas in other cases the properties of the magnetic field are modified by fluid motions. The tying of the fluid to the magnetic field lines also permits the propagation of what are known as hydro-magnetic or magnetohydrodynamic waves which have some similarity to sound waves but with a characteristic speed

$$c_H = (B^2/\mu_0\rho)^{\frac{1}{2}}, \tag{4.20}$$

instead of the sound speed

$$c_s = (\gamma P/\rho)^{\frac{1}{2}}, \tag{4.21}$$

in a gas, where γ is the ratio of the specific heat at constant pressure to the specific heat at constant volume in the gas. I now discuss each of these properties in slightly more detail. c_H is known as the hydromagnetic speed or the Alfvén speed.

Motion of a charged particle in a magnetic field

I consider first the motion of a single charged particle in a given electromagnetic field. I suppose that the particle has charge q; I reserve the symbol e for the magnitude of the charge on the electron, so that for an electron $q = -e$. The particle will have the equation of motion

$$m\frac{d\mathbf{v}}{dt} = q(\mathbf{E} + \mathbf{v} \times \mathbf{B}), \tag{4.22}$$

if its mass is m and its velocity is $\mathbf{v}$.* I consider first motion in fields which are constant in both space and time, $\mathbf{E}_0$ and $\mathbf{B}_0$. It can be shown that the motion of the particle has three components. If the electric field has a component, $E_{||}$, parallel to the magnetic field, the particle has accelerated motion in that direction. Thus

$$v_{||} = (q/m)E_{||}t + v_{||0}, \tag{4.23}$$

where $v_{||0}$ is the value of $v_{||}$ at $t=0$. There are two components of perpendicular motion. The particle drifts perpendicular to both the electric and magnetic field with the velocity

$$\mathbf{v}_D = \mathbf{E}_0 \times \mathbf{B}_0/B_0^2, \tag{4.24}$$

where B_0 is the magnitude of $\mathbf{B}_0$. Finally it moves in a circle about the magnetic field line. The frequency of the motion is

$$\omega_B = qB_0/m, \tag{4.25}$$

so that the period is $2\pi m/qB_0$. The velocity can have any value but the radius of the orbit scales with the value of the velocity.

These results are obtained in Appendix 5. Note that the directions of the motions described in (4.23) and (4.25) depend on the sign of q, whereas $\mathbf{v}_D$ is the same for charges of both signs. If there is no electric field (or more strictly if a frame can be chosen in which the electric field vanishes), motion along the magnetic field line is unaccelerated and $\mathbf{v}_D = \mathbf{0}$. This is usually a good approximation in astronomical problems. The particle then has a helical motion which is illustrated in fig. 22. This discussion of particle

---

* This equation is true if the speed of the particle is very small compared to c. It is easily generalised to the case of relativistic motion.

Fig. 22.   The helical motion of a charged particle along a magnetic field line.

motion in constant fields will be a good approximation to the true motion if two conditions are satisfied. In the case of zero **E** these are that the magnetic field must not change significantly in the time that it takes the particle to describe a circular orbit about it and that the distance over which there are spatial variations of the field must be large compared to the radius of the particle's orbit. This latter condition depends on the velocity or equivalently the energy of the particle. I shall say a little more shortly about what happens to a particle which moves in a spatially varying magnetic field.

If these two conditions are satisfied, I can say that the particle is effectively tied to the magnetic field lines, as it moves round them in a tight helix. If a gas is to be tied to the field lines, I need to ask what happens when I bring together a large assembly of particles. The new factor that enters is that the particles can interact with one another. These interactions are called collisions, if they have the effect of causing a significant change of direction in the motions of the particles. I can define a collision frequency ν_c whose reciprocal τ_c is the characteristic time between collisions for a particle. If the collision frequency is sufficiently large, the particle motions will become disordered and effectively decoupled from the magnetic field. In that case the fluid will not be tied to the field. If, in contrast, collisions are sufficiently rare, not only individual particles but the whole fluid will be effectively tied to the magnetic field lines. It is collisions which provide the electrical resistivity of the matter, because they can convert the ordered particle motions necessary for the flow of an electrical current into random motions. Equivalently the collisions prevent a fluid from having infinite electrical conductivity. In a fully ionised gas a good approximation to the value of the electrical conductivity is

$$\sigma = n_e e^2 \tau_c / m_e, \qquad (4.26)$$

where n_e and m_e are the number density and mass of the electron. As it can be shown that τ_c is proportional to $T^{3/2}/n_e$, the result $\sigma \propto T^{3/2}$ mentioned earlier is obtained. A brief discussion of electron/electron collisions can be found in Chapter 7. If the gas is not fully ionised, any neutral particles will not be directly affected by the electromagnetic field and collisions with ions and electrons will be required to keep them moving with the rest of the fluid.

Magnetic field in a moving medium

Earlier I discussed the decay of a magnetic field in a stationary medium. In obtaining equation (4.10) I omitted the term $\mathbf{v} \times \mathbf{B}$ from equation (4.6). If I now insert that term equation (4.8) is replaced by

$$\frac{\partial \mathbf{B}}{\partial t} = \mathrm{curl}\,(\mathbf{v} \times \mathbf{B}) + \frac{1}{\mu_0 \sigma}\, \nabla^2 \mathbf{B}. \tag{4.27}$$

This shows that there are two ways in which a magnetic field can be changed in a conducting fluid. It can be carried around by the fluid motions or it can decay as a result of finite electrical conductivity. If the second term is unimportant, the equation

$$\frac{\partial \mathbf{B}}{\partial t} = \mathrm{curl}\,(\mathbf{v} \times \mathbf{B}), \tag{4.28}$$

describes how the magnetic field is carried round by the motion of the fluid. Equation (4.28) can be shown to be equivalent to the statement that the magnetic flux through any element of fluid is conserved as the fluid moves; if a magnetic field $\mathbf{B}$ is perpendicular to an area A, the flux through the area is BA. I shall return to equation (4.27) shortly when I discuss the concept of dynamo maintenance of magnetic fields in a conducting fluid. I have already shown that, if the outer layers of the Sun were static, its magnetic field would have long since become current-free. As that is observed not to be the case, can the term $\mathrm{curl}\,(\mathbf{v} \times \mathbf{B})$ in equation (4.27) offset the field decay produced by the final term and lead to a field which is either time independent or which, on average, does not decay? The latter is what is required by the observations. I return to this point shortly.

Magnetohydrodynamic waves

In my discussion so far I have regarded the fluid velocity as something which is prescribed but I have also mentioned that in reality the magnetic field may constrain the motion of the fluid. In order to discuss this I need to write down an equation which connects the force exerted by the magnetic field with the fluid motions. This equation is

$$\rho \frac{\mathrm{d}\mathbf{v}}{\mathrm{d}t} = -\mathrm{grad}\,P + \mathbf{j} \times \mathbf{B} + \rho\,\mathrm{grad}\,\Phi. \tag{4.29}$$

In equation (4.29) I have assumed that the only forces acting are due to the gas pressure and the gravitational potential Φ and the magnetic force $\mathbf{j} \times \mathbf{B}$. This term is obtained by summing forces $q_i(\mathbf{v}_i \times \mathbf{B})$ acting on individual particles with charge q_i and velocity $\mathbf{v}_i$. This leads to the subject of magnetohydrodynamics or hydromagnetics. A full description of the system can then only be obtained if two further equations are used. These are the equation of continuity (or conservation of mass)

$$\frac{\mathrm{d}\rho}{\mathrm{d}t} + \rho\,\mathrm{div}\,\mathbf{v} = 0, \tag{4.30}$$

and an equation which relates the changes of P and ρ. The simplest form of this equation, which is frequently valid, is the adiabatic equation

$$\frac{1}{P}\frac{dP}{dt} = \frac{\gamma}{\rho}\frac{d\rho}{dt}. \tag{4.31}$$

Note that d/dt written in equations (4.29) to (4.31) differs from $\partial/\partial t$ in Maxwell's equations. $\partial/\partial t$ describes the rate of change with time of some quantity at a fixed point in space whereas d/dt is the rate of change with time following a fluid element moving with velocity $\mathbf{v}$. It can be shown that

$$\frac{d}{dt} = \frac{\partial}{\partial t} + \mathbf{v}.\text{grad}, \tag{4.32}$$

so that, for example,

$$\frac{d\rho}{dt} = \frac{\partial\rho}{\partial t} + \mathbf{v}.\text{grad}\,\rho. \tag{4.33}$$

It is possible to use the full set of magnetohydrodynamic equations to deduce the existence of the magnetohydrodynamic waves which I mentioned earlier. Once again I consider the simplest case of a medium of uniform density ρ_0 and pressure P_0 containing a uniform magnetic field $\mathbf{B}_0$. I also ignore the influence of the gravitational field and assume that σ is so large that (4.6) can be replaced by $\mathbf{E} + \mathbf{v} \times \mathbf{B} = \mathbf{0}$. In discussing wave propagation, I assume that all the physical quantities are perturbed by an amount which for any variable f has the form

$$f_1 \propto \exp i(\mathbf{k}.\mathbf{r} - \omega t) \equiv \exp i(k_x x + k_y y + k_z z - \omega t). \tag{4.34}$$

where $\mathbf{k}$ is the wave vector, ω is the frequency of the wave and it is the real part of (4.34) that is relevant. For any type of wave there is a relation, known as the dispersion relation between ω and $\mathbf{k}$. If there is no magnetic field, it is possible to show from equations (4.29) to (4.31) (ignoring gravity) that one type of wave can propagate, the sound wave, which has the dispersion relation

$$\omega^2 = k^2 c_s^2, \tag{4.35}$$

where $k = |\mathbf{k}|$. The sinusoidal form (4.34) travels through the fluid at the wave speed, $c_s = \omega/k$, which is called the phase velocity of the wave.

If a magnetic field is present, the magnetic force $\mathbf{j} \times \mathbf{B}$ couples equation (4.29) to the whole set of Maxwell equations. As a result the properties of sound waves are modified by the magnetic field and a new type of wave becomes possible. The magnetic field introduces a preferred direction into the system so that the properties of the waves depend on the direction in which they are travelling. In a uniform medium, sound travels equally strongly in all directions from its source; the same is not true of magnetohydrodynamic waves. It is now useful to write the magnetic field $\mathbf{B}_0 = B_0\mathbf{b}$, where $\mathbf{b}$ is then a unit vector in the direction of the field. It is shown in Appendix 6 that three types of wave can propagate. They are known as the Alfvén wave and the fast and slow magnetosonic (or magnetoacoustic) waves. The Alfvén wave has the dispersion relation

$$\omega^2 = (\mathbf{k.b})^2 c_H^2, \tag{4.36}$$

whereas the fast and slow magnetosonic waves have the coupled dispersion relation

$$\omega^4 - \omega^2 k^2 (c_s^2 + c_H^2) + k^2 (\mathbf{k.b})^2 c_s^2 c_H^2 = 0. \tag{4.37}$$

Various special cases are obvious. If the waves propagate along the field, $\mathbf{k.b} = k$, there are two waves with $\omega^2 = k^2 c_H^2$ and the sound wave $\omega^2 = k^2 c_s^2$, unaffected by the magnetic field. In contrast, for propagation perpendicular to the field only one wave survives with $\omega^2 = k^2 (c_s^2 + {}_H^2)$.

As I mentioned earlier and have just demonstrated explicitly the waves do not travel equally in all directions and this has important consequences when the magnetic field is strong in the sense that $c_H^2 \gg c_s^2$. When the opposite inequality is true, signals still travel essentially in all directions with speed c_s. When waves propagate anisotropically, it is necessary to introduce another wave velocity in addition to the phase velocity. This is the group velocity $\partial \omega / \partial \mathbf{k}$, with the three components $\partial \omega / \partial k_x$, $\partial \omega / \partial k_y$, $\partial \omega / \partial k_z$. This is the velocity with which the wave carries energy or information. This velocity, in general, differs in magnitude from the phase velocity and in direction from that of $\mathbf{k}$.* As a particular example the group velocity of the Alfvén wave is $c_H \mathbf{b}$ showing that, regardless of the direction in which it propagates, energy always travels along the field lines with speed c_H. Magnetic field lines are often thought of as analogous to stretched strings. Waves in a string with tension T and density ρ travel at a speed $(T/\rho)^{1/2}$. If the magnetic tension B^2/μ_0 is inserted into this formula, the hydromagnetic speed is obtained. The existence of hydromagnetic waves means that both energy and angular momentum can be transferred along magnetic field lines even without any flow of matter. This is important, for example, in the problem of star formation when the material which is going to form a star may be connected to other parts of the interstellar medium by a magnetic field. Thus, if the contracting protostar is rotating, as it contracts it will rotate more rapidly, if it conserves angular momentum. The magnetic field will however resist the consequent winding up of field lines and will transfer some of the angular momentum of the protostar along the field lines.

Magnetic braking

Another important effect can occur if a rotating magnetised star is losing mass, as for example the Sun does through the solar wind. The solar wind material is ionised and as a

---

* An analogous case is that of electromagnetic waves in a dielectric where the phase velocity can exceed c but the group velocity is always less than c, thus reflecting the principle of relativity.

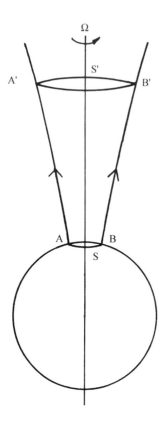

Fig. 23. Magnetic braking. Material leaving a star through the surface S is forced to co-rotate with the star as it moves along diverging magnetic field lines to surface S'. As a result it carries away angular momentum which reduces the star's angular velocity.

result the particles must flow along the magnetic field lines. As I have mentioned earlier this means either that the magnetic field controls the flow or that the outflowing matter carries the magnetic field lines with it. To a first approximation it proves true to say that, if the energy density of the magnetic field ($B^2/2\mu_0$) is greater than the kinetic energy of the matter ($\rho v^2/2$, where v is the outflow velocity), the matter is tied to the magnetic field lines, whereas, if the kinetic energy of the matter is greater, the magnetic field lines are carried about by the matter.

This leads to a concept known as magnetic braking which is illustrated in fig. 23. Suppose that a magnetic field structure is rotating as a result of the rotation of the star in which it is produced. Material leaving the surface of the star at point A travels along the field line shown until it reaches A', where its kinetic energy density equals the magnetic field energy density. This is called the Alfvén point in the outflow. After that it is able to continue to move outwards carrying the magnetic field with it. It is easy to see how the relative importance of the kinetic and magnetic energy densities changes so as to make this possible.

Assume that, as is observed in the case of the Sun, the outflowing material is moving at well above escape velocity, so that I can ignore its slowing down due to the gravitational attraction of the star. Then suppose field lines at A define a surface S as shown and those at A' a corresponding surface S'. The flow of mass through S must equal that through S', which means

$$\rho v S = \rho' v' S'$$

or

$$\rho S = \rho' S', \tag{4.38}$$

since $v = v'$. The magnetic flux through S also equals that through S', which gives

$$BS = B' S'. \tag{4.39}$$

These equations can be combined to give

$$\frac{\rho' v'^2}{B'^2/2\mu_0} = \frac{\rho' v^2}{B'^2/2\mu_0} = \frac{\rho v^2}{B^2/2\mu_0} \frac{S'}{S}, \tag{4.40}$$

which shows that as the field lines move apart, which they must in general do as the flux from the star is spread out over a larger area, the importance of the kinetic energy increases relative to the magnetic energy so that there will be a radius where they become equal, if the magnetic energy was initially larger.

Now consider the angular momentum possessed by the outflowing material at A'. The magnetic field is rotating rigidly with the angular velocity Ω of the star at A and the outflowing matter must share this rotation until it possesses a high enough relative energy density to control the structure of the field. The angular momentum per unit mass at A is Ωd^2, where d is the distance from the rotation axis, but by the time the material is at A' it has the larger value $\Omega d'^2$, where d' is the new distance from the rotation axis. Where has this angular momentum come from? It can only have come from the rotation of the star itself. Thus the outflowing material, while it is channelled by the magnetic field, carries away angular momentum and slows down the stellar rotation. This is known as magnetic braking and it is believed to play an important role in early stellar evolution. A full treatment is more complicated than this brief discussion but the qualitative result is unchanged.

Magnetic reconnection

As I have explained in Chapter 2 a large amount of energy is released by localised solar flares. In addition some explanation is required of the very high kinetic temperature of the solar corona. It is believed that the dissipation of magnetic energy is important in both of these processes. The phenomenon known as magnetic reconnection is thought to be relevant in both cases. The basic idea of magnetic reconnection is illustrated in fig. 24. Suppose two oppositely directed magnetic field lines are brought together by fluid motions. As the field lines approach, the gradient of the magnetic field will

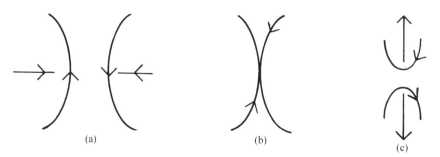

Fig. 24. Magnetic field line reconnection. In (a) two oppositely directed field lines are moving towards one another. In (b) they make contact and in (c) they have reconnected and are separating.

become extremely strong in the region between them, where there will be what is known as a neutral point; a point where the magnetic field vanishes. As the field gradient increases the resistive dissipation produced by the term $(1/\mu_0\sigma)\nabla^2\mathbf{B}$ in equation (4.27) will become important even if the conductivity σ is very large. The effect of dissipation is to cause a reconnection of the magnetic field lines which then move apart in a direction perpendicular to their original motion. This is also shown in fig. 24. As a consequence, magnetic energy has been released in the neighbourhood of the reconnection point.

Although the basic idea of magnetic reconnection and of its possible significance in solar flares dates back at least to the 1950s, the fine details of the process and of its application are still a subject of active research today. In fig. 24, I have only shown the behaviour of one magnetic field line, whereas the field configuration must be continuous. Such a possible continuous configuration is shown in fig. 25. Here fluid motions down the y axis push together oppositely directed field lines at the origin. As a result of field line reconnection, fluid is squirted out in the x direction together with the reconnected field lines. Note that the reconnection point will not in general be a true neutral point of the magnetic field as there can be a field in the z direction as well. The field gradients will remain high even in the presence of such an additional component, B_z, and reconnection will still occur. It is possible to envisage a situation in which a configuration like that of fig. 25 continues for a long time as a result of additional field lines being carried into the reconnection region (steady-state reconnection) but for violent events such as flares a large energy release in a short time is required.

Magnetic fields and convection

Another important phenomenon in the outer layers of the Sun and other stars is the interaction of magnetic fields with convection. As I shall discuss later, the convective motions are involved in the whole process of dynamo maintenance of stellar magnetic fields. Before discussing that, I consider some related points. It is possible to discuss the

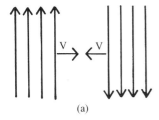

(a)

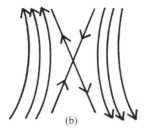

(b)

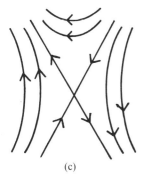

(c)

Fig. 25. A process of steady-state reconnection, showing successive field lines approaching and reconnecting.

effect on an initially very simple magnetic field of an equally simple fluid motion by solving equation (4.27) numerically. This is illustrated in fig. 26. A uniform magnetic field crosses a square box and it is influenced by an eddy type motion circling the centre of the box. What precisely happens must depend on the value of the magnetic diffusivity $(1/\mu_0\sigma)$ or more precisely on the ratio of the eddy turnover time, L/v, to the resistive decay time $\mu_0\sigma L^2$; the reciprocal of this ratio, $Lv\mu_0\sigma$, is called the magnetic Reynolds number for the flow. L is the length of the side of the box and v is a typical value of the velocity. If the magnetic Reynolds number is very low the field is unaffected by the motions; if the Reynolds number is very high, it is wound up many times before any dissipation occurs.

The second diagram of fig. 26 shows what happens for an intermediate value of the Reynolds number. The magnetic field is largely swept clear from the centre of the eddy and is concentrated in flux ropes at the edge. The boundary conditions for this particular calculation are such that the total flux entering the top and leaving the bottom of the box must be conserved. The idea of the expulsion of magnetic flux by eddies is a very important one, particularly when it is allied with the idea of magnetic buoyancy which I introduced earlier in the chapter. If convection in a star can be represented as a system of eddies, the magnetic field in the convection zone may be concentrated into buoyant flux ropes which rise towards the surface and this in turn may be related to the appearance of

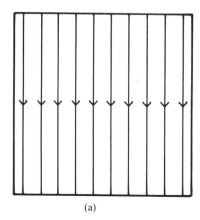 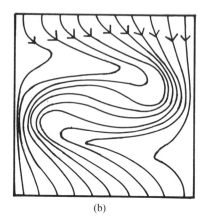

(a)   (b)

Fig. 26.   Effect of fluid motions on a magnetic field. A simple vortex type motion moves the straight magnetic field lines (marked with arrows) of (a) into those shown in (b). Where the field lines are closer together the magnetic induction is increased substantially and the average value of the magnetic induction is also increased.

sunspots. It may also be related to the concentration of the surface magnetic field of the Sun away from spots in thin filaments.

Convection may involve eddies of many different length scales. Some may be large enough to affect the overall structure of the magnetic field in the manner which I have just described. Others may themselves be influenced by a strong magnetic field. At the surface of the Sun the visible granulation which I have already described in Chapter 2 is evidence for convection near the surface. The length scale of this granulation is much smaller than the size of a large sunspot and it is possible to see that granulation is suppressed in a sunspot. This suggests a consideration of a different problem which is that of the interference of a strong magnetic field with convection. In the absence of a magnetic field convection occurs in a gas of ratio of specific heats γ if the ratio of the temperature gradient to the pressure gradient satisfies the relation

$$\frac{P}{T}\frac{\mathrm{d}T}{\mathrm{d}P} > \frac{\gamma - 1}{\gamma},$$ (4.41)

as has been discussed in Chapter 3. If a vertical magnetic field of strength B threads the fluid, it can be shown that this criterion is modified to

$$\frac{P}{T}\frac{\mathrm{d}T}{\mathrm{d}P} > \frac{\gamma - 1}{\gamma} + \frac{B^2}{B^2 + \gamma P},$$ (4.42)

indicating that a strong enough magnetic field can prevent convection from occurring and that a weaker field can interfere with convection. The magnetic field cannot prevent motions which are oscillatory up and down the field lines but these are likely to be less efficient at carrying energy than those that turn over. That sunspots are cooler than their surroundings can then be associated with a reduction of efficiency of convection inside the magnetic flux tubes of the spots, as has already been suggested in Chapter 2.

Dynamo action

As I have said earlier many cosmic magnetic fields would have decayed by now if they had not been continuously regenerated. The process of regeneration is believed to be the interaction of fluid motions with a magnetic field as described by equation (4.27). In the case of the outer regions of the Sun there are two types of motion. There is the large scale motion of solar rotation which is non-uniform as is clear from the observed differential rotation of the solar surface. There are also the small scale convective motions. It is believed that the differential rotation and convection are responsible for the dynamo maintenance of magnetic fields in the Sun and other stars.

A full discussion of the possible dynamo maintenance of magnetic fields is very complex. The reason for this is that equation (4.27) is coupled with the fluid dynamic equations through which the velocity $\mathbf{v}$ is determined and where the magnetic field influences the value of $\mathbf{v}$ through the Lorentz force $\mathbf{j} \times \mathbf{B}$. Most work on dynamo theory has sidestepped this difficulty by discussing *kinematic dynamos*. In this treatment the fluid velocity is prescribed and it is asked what properties this velocity must have if it is to produce dynamo action. It can then be considered whether the actual velocities are likely to possess the required properties. There would be no point in trying to solve the full dynamical problem if it could be shown that under no circumstances was a dynamo possible.

For a long time the problem was that of demonstrating that dynamos might exist. It was originally conjectured by J J Larmor in 1919 that some cosmic magnetic fields might be maintained against ohmic decay by fluid motions. The first serious treatment was by T G Cowling in 1934 when he proved an anti-dynamo theorem. In order to obtain a soluble problem he studied configurations which had an axis of symmetry and he showed that axisymmetric fluid motions could not maintain a time-independent magnetic field. For a significant time after this there was no real progress. Relaxation of Cowling's assumptions meant that further progress might rely on numerical work and powerful enough computers did not exist even for a serious attempt at the problem until the 1950s. Then in 1958 there were two independent demonstrations of the possibility of dynamo action by A Herzenberg and G Backus involving non-axisymmetric motions.

The main difficulty in establishing dynamo action can be explained quite simply. Consider cylindrical polar coordinates $(\tilde{\omega}, \phi, z)$. The field B_ϕ is called a *toroidal field* and the other two components are called a *poloidal field*. Suppose that initially there is only a poloidal field but that the medium is then put into a state of non-uniform rotation about the z axis. This motion pulls out magnetic field lines in the toroidal direction and produces a toroidal field (fig. 27). It is clear, similarly, that motions in the poloidal direction can produce a poloidal field out of a toroidal field. The crucial question is whether, when ohmic dissipation is included, a situation can be reached in which the

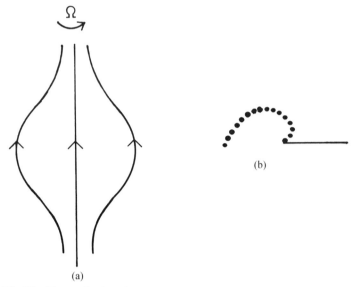

Fig. 27. The production of a toroidal magnetic field by differential rotation. (a) shows the initial poloidal field configuration. (b) shows the projection of a single field line on the midplane. The straight line is the projection before rotation starts and the dots show the projection after some rotation has occurred. The departure from a straight line shows that a toroidal field has been produced.

regeneration of each component of the field offsets its decay. What Cowling demonstrated was that with axisymmetry the decay always wins. In contrast, Backus and Herzenberg showed that it was in principle possible for non-axisymmetric motions to maintain fields.

Once it had been established that dynamo action was possible in principle, the serious development of the subject started. I shall give a brief description of one approach known as *mean field electrodynamics*. In this method the magnetic induction and the fluid velocity are both supposed to be divided into mean vectors and small scale fluctuating vectors. Thus I write

$$\mathbf{B} = \mathbf{B}_0 + \mathbf{b}, \quad \mathbf{v} = \mathbf{v}_0 + \mathbf{u}, \tag{4.43}$$

where the mean field and velocity, $\mathbf{B}_0$ and $\mathbf{v}_0$, are assumed to be slowly varying in space and time whereas the average values of $\mathbf{b}$ and $\mathbf{u}$, which are more rapidly varying, vanish; $\mathbf{b}$ must not be confused with the unit vector $\mathbf{b}$ introduced earlier in the chapter. As I am studying the kinematic dynamo problem, I assume that $\mathbf{u}$ is specified while $\mathbf{b}$ is to be determined. If (4.43) is now inserted into equation (4.27), this can be separated into two equations for mean and fluctuating parts. Thus

$$\frac{\partial \mathbf{B}_0}{\partial t} = \text{curl}(\mathbf{v}_0 \times \mathbf{B}_0) + \text{curl}\langle \mathbf{u} \times \mathbf{b}\rangle + \eta \nabla^2 \mathbf{B}_0 \tag{4.44}$$

$$\frac{\partial \mathbf{b}}{\partial t} = \mathrm{curl}(\mathbf{v}_0 \times \mathbf{b}) + \mathrm{curl}(\mathbf{u} \times \mathbf{B}_0) + \mathrm{curl}(\mathbf{u} \times \mathbf{b} - \langle \mathbf{u} \times \mathbf{b} \rangle) + \eta \nabla^2 \mathbf{B}, \qquad (4.45)$$

where $\eta \equiv 1/\mu_0 \sigma$ is the resistivity and $\langle \; \rangle$ denotes an average value.

To make any progress with a discussion of the evolution of the mean magnetic induction $\mathbf{B}_0$ an expression is now required for $\mathrm{curl}(\mathbf{u} \times \mathbf{b})$. As is explained in slightly more detail in Appendix 7, equation (4.45) indicates that, if $\mathbf{b}$ were initially zero, its subsequent value would be linearly dependent on $\mathbf{B}_0$. As $\mathbf{u}$ is prescribed, $\mathbf{u} \times \mathbf{b}$ must also depend linearly on $\mathbf{B}_0$. A possible form for $\langle \mathbf{u} \times \mathbf{b} \rangle$ is

$$\langle \mathbf{u} \times \mathbf{b} \rangle = \alpha \mathbf{B}_0 - \beta \, \mathrm{curl} \, \mathbf{B}_0. \qquad (4.46)$$

The appropriate expressions for α and β depend on the form of the prescribed velocity $\mathbf{u}$ and on the zero order magnetic field $\mathbf{B}_0$ and it is well beyond the scope of the present book to discuss their derivation in detail. If the fluctuating velocity is one of turbulent convection, appropriate expressions for α and β can be shown to be

$$\alpha = \tau_{\mathrm{conv}}(\mathbf{u}.\mathrm{curl}\,\mathbf{u})/3, \qquad (4.47)$$
$$\beta = \mathbf{u}^2 \tau_{\mathrm{conv}}/2, \qquad (4.48)$$

where τ_{conv} is the characteristic convective turnover time.

Equations (4.46) to (4.48) can then be inserted into (4.44) to give,

$$\frac{\partial \mathbf{B}_0}{\partial t} = -\mathrm{curl}(\eta + \beta)\mathrm{curl}\mathbf{B}_0 + \mathrm{curl}(\mathbf{v}_0 \times \mathbf{B}_0) + \mathrm{curl}\,\alpha \mathbf{B}_0. \qquad (4.49)$$

In going from (4.44) to (4.46) it has been recognised that in reality η, as well as β, is not a constant If η is a constant

$$-\eta \, \mathrm{curl} \, \mathrm{curl} \, \mathbf{B}_0 \equiv \eta \nabla^2 \mathbf{B}_0. \qquad (4.50)$$

It can be seen that β is an addition to η; η is the normal resistivity and β is known as the *turbulent resistivity*. The turbulent resistivity is frequently much larger than the normal resistivity, which means that in the absence of the final term in equation (4.49) the decay of magnetic fields would be enhanced. The final term, which is known as the α-effect, is crucial for dynamo action.

The study of the dynamo maintenance of magnetic fields is a very active area of research at the present time. Many model problems of kinematic dynamics in a simple geometry have confirmed that dynamos can work. A full understanding of the problem requires a study of the geometry of spheres and spherical shells which are relevant in astronomy. In addition the kinematic discussion must be replaced by the complete dynamical discussion in which the origin of the fluid motions and their interaction with the magnetic field is fully included. There seems no doubt that the dynamo process exists and is important but it is likely to be some time before it is fully understood.

Summary of Chapter 4

Magnetic fields in highly conducting fluids decay slowly and can often be regarded as essentially permanent, whereas electric fields, other than those which arise from magnetic fields when a frame of reference is changed, are usually transitory. In some stars the decay time of a magnetic field is longer than the main sequence lifetime of the star, but this is not true of the magnetic field in the outer layers of the Sun. There could in principle be a strong long-lived field in the centre of the Sun but a variety of observations and theoretical ideas indicates that this is highly unlikely. The observed surface magnetic field is in any case not a simple extension of a long-lived internal field and it must be maintained by what is known as a dynamo process.

A single charged particle moving in a magnetic field describes a helical orbit around a field line. In most astronomical contexts the radius of the orbit is much smaller than the distance in which the field changes significantly. If a large number of particles is present, the fluid is effectively tied to the field provided that the frequency of particle/particle collisions is very low compared with the particle gyration frequency in the field. As a result, a strong field constrains the motion of the fluid, whereas very energetic fluid motions carry the magnetic field around with them. Field/fluid interactions cause the magnetic field lines to behave like stretched strings along which waves known as hydromagnetic or magneto-hydrodynamic waves can propagate. Other important consequences of the interaction between fluids and magnetic fields include the instability known as magnetic buoyancy and the process of magnetic braking. An isolated tube of magnetic flux sitting in a conducting fluid is in unstable equilibrium, because equilibrium requires the density inside the tube to be less than that outside and a buoyancy force causes the flux tube to rise. Magnetic braking occurs if a rotating magnetised object, such as a star, loses mass which flows along diverging field lines. If the field is strong enough to control the flow, the expelled mass gains angular momentum and the underlying object is slowed down.

Although the structure of magnetic fields does not usually change rapidly in a fluid of high conductivity, this does not remain the case if oppositely directed field lines are brought together by fluid motions and meet in what is known as a magnetic neutral point or neutral line. In such a case a rapid change in magnetic topology and a large release of magnetic energy can occur. The process known as magnetic reconnection is believed to be important in solar flares and possibly in the heating of the solar corona, and elsewhere when a rapid local release of energy is observed.

In stars like the Sun the main motions which interact with magnetic fields are those in the outer convection zone. They and the observed solar differential rotation are believed to be responsible for the dynamo maintenance of magnetic fields. Convective motions occur on many length scales. The large scale motions can move the magnetic field lines around and can produce the tubes of magnetic flux which subsequently become buoyant. In contrast, inside such a flux tube the magnetic field may interfere with the convection and this is responsible for the low temperature of sunspots compared with the neighbouring photosphere.

Although the dynamo maintenance of magnetic fields was first suggested by Larmor in 1919, it was almost 40 years before it was demonstrated that it was possible. In 1934 Cowling showed that axisymmetric motions could not maintain a field and there was then a long delay before the proof that there existed non-axisymmetric motions that would do so. From that proof that prescribed motions can maintain a steady field, there is a long step to the demonstration that naturally occurring non-axisymmetric motions in a differentially rotating convection zone, where the motions are influenced by the field rather than being prescribed, will maintain a quasi-steady field. Although it cannot be claimed that the problem is fully solved, it is generally accepted that dynamos do operate in the outer convection zones of late-type stars as well as the liquid core of the Earth.

5   The active Sun

Introduction

I have been preparing for this chapter in each of the three previous chapters. In Chapter 2 I gave a brief account of the relevant observations and in Chapter 3 I discussed how the overall internal structure of the Sun can in principle be calculated and in particular explained that the Sun has a deep outer convection zone. Finally in Chapter 4 I have described the interactions between magnetic fields and particles and fluids which are relevant if a theoretical understanding is to be obtained of the complex processes which are observed in the outer layers of the Sun. My aim in this chapter is to discuss the extent to which such an understanding has been obtained. The subject is so complicated that I shall only be able to present a very much oversimplified view of what occurs.

The solar dynamo

I have explained in Chapter 4 that a suitably designed set of non-axisymmetric fluid motions can maintain a magnetic field against the effects of ohmic dissipation. The simplest discussion involves the maintenance of a time-independent magnetic field by prescribed fluid motions. This is both more and less than is required in the Sun. The observed magnetic field is not time-independent so what is required is that in some average sense the field should be maintained. On the other hand, the magnetic field is observed to have such a strength that it should react back on the fluid motions. This means that the properties of the convection, the differential rotation of the Sun and of the magnetic field should be determined self-consistently. Thus, for example, it is to be hoped that a dynamo process acting in the convection zone of an initially uniformly rotating Sun would lead to the observed equatorial acceleration at the surface.

Out of such a calculation there should emerge a characteristic timescale of about 22 years if there is to be a natural explanation of the solar cycle. I have explained in Chapter 4 that it is believed that fluid motions acting on the magnetic field in the deep convection zone are thought to confine it in bundles of flux and that these flux tubes rise as a result of magnetic buoyancy until they break through the solar photosphere and become visible as bipolar spot pairs. A successful account of the solar dynamo should explain why the spots appear at high (but not too high) latitudes early in the solar cycle but at lower latitudes as the cycle progresses. It is the rôle of the dynamo theory to provide a general structure to the magnetic field which exists close to the solar surface, but not to explain the detailed magnetic activity which then results. If such an understanding could be obtained of the solar dynamo, it should in principle then be possible to make predictions for the properties of the dynamo in other stars with different surface temperatures, luminosities, rotation rates and depth of outer convection zone, as a prelude to obtaining an understanding of the properties of other active stars.

Consider first what is known about the solar dynamo. I have already shown a butterfly diagram of the distribution of sunspots in Chapter 2. A more up to date version is shown in fig. 28. It can be seen that the sunspot cycle is only approximately periodic in two senses; both the total number of sunspots (or the area covered by them) and the length of the cycle in years are variable. I have already mentioned the Maunder minimum in the 17th Century in which there were very few sunspots observed. A recent careful investigation has shown that this reported lack of sunspots could not have been caused by observational selection. Observers at Paris studied the Sun just as regularly in the period of the Maunder minimum as they did in the more active period which followed it. The sunspots which were observed during the Maunder minimum were all relatively close to the equator and almost all of them were in the southern solar hemisphere. There is some suggestion from the numbers that an underlying solar cycle persisted during the minimum but the statistics are rather poor. It is obviously of interest to know whether the Maunder minimum was a very rare event or whether such events have occurred regularly in past history. Unfortunately it occurred so soon after the start of regular sunspot observations that this is a question that cannot be answered directly.

Fortunately there are other indicators of past sunspot activity. The Earth's atmosphere is constantly being bombarded by galactic cosmic rays, which produce the radioactive isotopes ^{14}C and ^{10}Be, amongst others. These isotopes are subsequently taken up by living organisms, the beryllium quite rapidly and the carbon after a typical delay of about 30 yr. Once they are in the organisms they decay and the normal method of ^{14}C dating compares the relative amount of ^{14}C to ^{12}C in the organism or something made from it, with what was originally present to estimate how old the object is. This method assumes a known rate of production of ^{14}C. Here we want something different. When the Sun is very active it is surrounded by an increased magnetic field and this reduces the flux of galactic cosmic rays at Earth, because of the difficulty which charged particles have in crossing field lines. This means that the production of ^{14}C should not be constant in time; it should be greater at sunspot minimum that at sunspot maximum. There is an additional effect on a longer timescale because of variations in the geomagnetic field. Such variations in the production of ^{14}C affects the accuracy of ^{14}C dating.

Studies of ^{14}C and ^{10}Be in the polar ice-caps have shown clearly that their production rates were higher at the time of the Maunder minimum than at present. Moreover studies

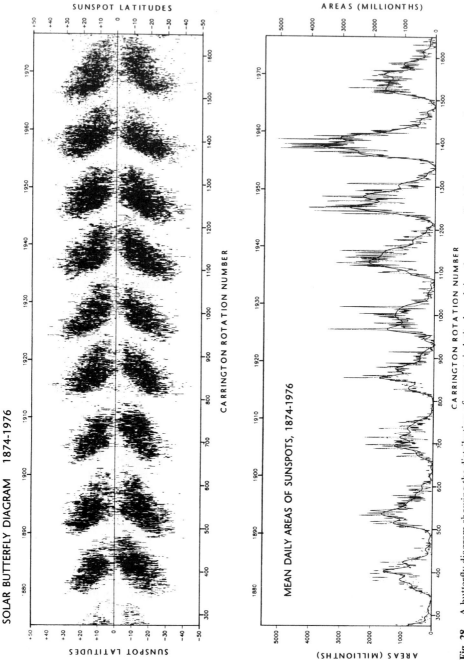

Fig. 28. A butterfly diagram showing the distribution of sunspots in latitude and time (upper panel) with the corresponding area covered by sunspots (lower panel) (from the Greenwich photoheliographic observations, Yallop, B D & Hohenkirk, C Y, *Solar Physics*, **68**, 303, 1980).

of deeper layers in the ice have shown that there were other maxima in ^{14}C and ^{10}Be production in the past with something like a Maunder minimum occurring typically every 200 years but with a significant variation in the spacing. ^{14}C has been studied for the past 9000 years and the record shows a smooth change in the rate of production attributed to changes in the geomagnetic field upon which are superimposed many fluctuations which could have been caused by Maunder minimum type events. The conclusion drawn is that the Maunder minimum was not a one-off very rare occurrence but was something which happens quite frequently and which should be explained by the dynamo theory.

There is a further controversial observation relating to the state of the Sun during the time of the Maunder minimum. Observations made at Paris suggest that the diameter of the Sun was increased by a fraction of order 2×10^{-3} at the time of the minimum. In addition there was an approximately 10 year modulation in the radius which suggests that there was dynamo action in the solar interior, which was not becoming apparent at the surface. There is dispute about the interpretation of these observations and some astronomers argue that the changes in diameter are spurious and that they can be explained away by a combination of observing errors, instrumental imperfections and variations in properties of the Earth's atmosphere, through which the Sun was observed. Whether or not the quoted change in radius is accurate, it has to be recognised that it would be surprising if there were no such change of radius at all. The radius of the Sun is determined by the forces acting in the interior and by the mechanism of energy transport. The fine detail must include the magnetic pressure and the turbulent pressure due to the convective motions. In the Maunder minimum, both the pattern of convection and the structure of the magnetic field must have been different from what they are today. In addition, although there was a small reduction in solar luminosity, the energy release in the solar core cannot have changed. This implies that some additional energy must have been stored in the solar outer layers. Because the internal properties of the magnetic field and the convection are not known, it is not easy to see what sign the change in radius should have.

Site of the dynamo

There is a general agreement today that the region in the Sun where the principal dynamo action occurs is the so-called overshoot region at the base of the convection zone. Convection cannot cease abruptly where the criterion for the occurrence of convection ceases to be satisfied. Convective elements are moving and they must come to rest in a region which is not unstable to convection. This is known as convective overshooting. The surface rotation velocity of the Sun exhibits an equatorial acceleration. Helioseismological studies have shown that this dependence of angular velocity with latitude persists through the convection zone but that there is then a very rapid change to almost uniform rotation in the overshoot region. This means that there is a very strong velocity shear which might be crucial in driving the dynamo. This leads to the natural deduction that such an overshoot layer is also the site of the dynamo in other active stars. The general idea then is that the magnetic field which is generated by the dynamo is pushed together into flux tubes and that these flux tubes then rise rapidly through the convection zone under the effect of magnetic buoyancy until they break through the surface as bipolar spot pairs. Note that the overshoot layer is not believed to be extremely thin. It has an estimated thickness of a few times 10^7 m compared to a total solar radius of 7×10^8 m.

The statement that the principal source of the dynamo is at or below the base of the convection zone does not mean that the interaction between magnetic field, rotation and convection in the main part of the convection zone is totally unimportant. It must be recognised that the picture which I have earlier presented of magnetic buoyancy is oversimplified. Magnetic field lines do not have ends and it is therefore not possible to consider the buoyant rise of the section of flux tube shown in fig. 21 without asking how the ends of the field lines are connected. Only for the case of a purely toroidal magnetic field, where the tube of flux is closed rather like a smoke ring, is the concept of buoyancy completely clear. In general, the buoyant rise of a section of a flux tube might be resisted by the tension in the field lines which have to be extended by the buoyant motion. In addition any buoyant flux tube will not be able to rise through an otherwise static medium. The flux tube must interact with the convective motions and with the differential rotation. Some recent calculations show flux tubes being driven down rather than up by such motions. Having said that, some flux tubes must break through the solar surface to produce the sunspots and the associated magnetic activity. It also has to be recognised that much of the magnetic flux which is present in the solar convection zone and which passes through the surface of the Sun is not in the large flux tubes associated with sunspots. As I have explained in Chapter 2 there are much smaller concentrations of field.

Dynamo models

It is not possible at the moment to perform fully-consistent hydrodynamic calculations of the development of a dynamo in the outer layers of a star. It is necessary to have some approximations in the model. Calculations have been performed of convection in a rapidly rotating spherical shell in which the lower surface is heated. The fluid is taken to be incompressible apart from allowing for the effect of thermal expansion in encouraging convection. A small magnetic field is introduced and it is amplified by a dynamo type process. In such calculations differential rotation is produced but the waves of dynamo activity propagate towards the poles rather than towards the equator. It is clear that these simple models cannot account for the drift of magnetic activity towards the equator observed in the Sun. Calculations have been made for different values of the angular velocity and of the resistivity (or of a parameter related to it known as the magnetic diffusivity). It is found that, in a limited parameter range, a cyclic dynamo process is obtained. These calculations cannot explain the solar cycle but they contain some clues to the required behaviour and, as computers continue to get more powerful, it may be possible to obtain results which are nearer to observations.

An alternative approach is to study model problems which make no attempt to have the correct geometrical properties for a solar or stellar dynamo. These problems may however provide useful clues to behaviour which may occur in real solar or stellar dynamos. One particular property of some of the models is the occurrence of chaotic behaviour. As has already been explained and shown in figures, the solar dynamo is only very approximately periodic in the sense that both the amplitude and period of the activity vary from cycle to cycle. In addition the Maunder minimum was a time in which there was a dramatic departure from the normal approximately regular behaviour. The observed properties are similar to what one might expect from a non-linear dynamo exhibiting chaotic behaviour.

After these few remarks about the solar dynamo, I shall assume that there is a mechanism for regeneration of magnetic flux, which is situated deep in the solar convection zone and I shall turn to the visible manifestations of magnetic activity, which are all believed to be related to the existence of the dynamo. Here there are those aspects of the activity which are always present though variable in intensity and those which are more strictly intermittent. In the former category I place the chromosphere, transition region, corona and solar wind and in the latter category, flares, prominences and sunspots. I discuss the former category first.

The solar wind

It was recognised by E N Parker that a solar wind was a necessary consequence of the existence of the observed high temperature solar corona. The argument, which he presented and which I summarise below, was for the case of a spherically symmetric corona. It was already known from eclipse observations of the corona that departures from spherical symmetry were very important and, in particular, that the shape of the corona varied through the sunspot cycle. The associated spatial variation in the coronal temperature has become more apparent in the X-ray observations showing coronal hot spots and holes. Nevertheless, the spherical explanation for the existence of the solar wind remains qualitatively correct.

The argument for the existence of the solar wind is as follows. Suppose the hot corona sits in static equilibrium on top of the Sun. In this case the pressure gradient in the corona must be balanced by the gravitational attraction of the Sun. I can write equation (3.1) in the form

$$dP/dr = -GM_\odot \rho/r^2,   \tag{5.1}$$

where because the total mass in the corona is very small I can replace a variable M by $M_\odot$ on the right hand side of (5.1). I also have the relations

$$P = nkT_{\text{kin}}, \quad \rho = nm,   \tag{5.2}$$

where n is the number of particles per unit volume and m is the average particle mass, which is determined by the chemical composition and ionisation state of the corona. T_{kin} is the kinetic temperature of the coronal material. It is not a true thermodynamic temperature as conditions in the corona are very far from thermodynamic equilibrium.

One further equation is required, which relates to the flow of heat in the corona. Conduction is likely to be much more important than radiation. If κ is the coefficient of heat conduction, its form for an ionised gas is

$$\kappa = \kappa_0 T_{kin}^{5/2}, \tag{5.3}$$

where κ_0 is a constant. Assuming that there is no internal release of heat in the corona, the outward flow of heat, L_{cond}, must be a constant, which is independent of radius. Thus

$$L_{cond} = -4\pi r^2 \kappa_0 T_{kin}^{5/2} dT_{kin}/dr = \text{constant}. \tag{5.4}$$

Equation (5.4) can be integrated to obtain

$$T_{kin}/T_c = (r_c/r)^{2/7}, \tag{5.5}$$

where r_c and T_c are the radius and temperature at some point in the corona.

If equations (5.1), (5.2) and (5.5) are combined, equation (5.1) can then be solved to give P and n (or ρ) as a function of r. There is no analytical solution but a numerical solution showed that the corona should have a kinetic temperature of about 4.4×10^5 K and a particle density of about 4×10^8 m^{-3} in the vicinity of the Earth. Before the launch of the first spacecraft, there was no clear contradiction with the properties of the Earth's neighbourhood. What Parker pointed out was that there was a clear problem, if the solution of equation (5.1) was extended to the edge of the solar system. It is possible to show that at large values of r, the value of P tends to a constant value so that from (5.2) and (5.5) $\rho \propto r^{2/7}$. The numerical value of this constant pressure is higher than any estimate of the pressure of the interstellar medium near to the Sun or anywhere in the Galaxy. This means that a static model of the corona does not make sense. Parker discovered this property of the static corona model and decided that the corona must expand. If the corona is expanding, there is an outward flow of energy due to the mass motions and this modifies the temperature and density distribution as a function of radius.

Solar wind equations

If the material in the corona moves outward with a velocity v_r in the radial direction, equation (5.1) becomes (using (5.2))

$$nmv_r \frac{dv_r}{dr} = -\frac{d}{dr}(nkT_{kin}) - \frac{GnmM_\odot}{r^2}. \tag{5.6}$$

In addition, mass conservation in the outflowing material requires

$$nr^2 v_r = \text{constant}. \tag{5.7}$$

It is also necessary to modify the thermal conduction equation to allow for the outward flow of kinetic energy. The resulting equations can only be solved numerically, although some of the general properties of the solution can be found without integration. If the following substitutions are made,

$$\xi = r/a, \tag{5.8}$$
$$\tau = T_{\text{kin}}/T_0, \tag{5.9}$$
$$\lambda = GmM_\odot/akT_0, \tag{5.10}$$
$$\psi = mv_r^2/kT_0, \tag{5.11}$$

where a is the radius at the base of the corona and T_0 is the value of T_{kin} there, equations (5.6) and (5.7) can be combined to give

$$\frac{\mathrm{d}\Psi}{\mathrm{d}\xi}\left[1 - \frac{\tau}{\psi}\right] = -2\xi^2 \frac{\mathrm{d}}{\mathrm{d}\xi}\left[\frac{\tau}{\xi^2}\right] - \frac{2\lambda}{\xi^2}. \tag{5.12}$$

In Parker's original discussion he did not use an equation for T_{kin}. Instead he argued that the general character of the results could be obtained by considering an isothermal corona, $T_{\text{kin}} \equiv T_0$. His assumption was justified by more detailed discussion. This means he has $\tau = 1$. With this value of τ and noting that $\lambda = v_{\text{esc}}^2/v_{\text{th}}^2$, where v_{esc} is the escape velocity from the Sun and $v_{\text{th}}^2 = 2kT_0/m$, it can be seen that there must be a value of ξ at which the right hand side of (5.12) vanishes. The left hand side must also vanish. The only way this can happen and give a velocity above the escape velocity at large distance is to have $\psi = \tau$ there, or $\psi = 1$ in Parker's original model. Note that $\psi = 1$ is equivalent to $v_r = v_{\text{th}}$, and this critical point is loosely called a sonic point. Solutions passing through the sonic point can attain a high velocity far from the Sun. Figure 29 is based on Parker's original calculations. As I have said above, non-isothermal solutions have the same general character.

The model of the solar wind which I have discussed above is spherically symmetrical and time-independent. Its properties really vary with solar activity both through the sunspot cycle and as a result of solar flares. In addition, the outflow of matter is closely related to the structure of the solar magnetic field. In Chapter 7, I shall explain that the magnetic field controls the behaviour of the solar wind close enough to the solar surface but that eventually the solar wind drags the magnetic field lines through interplanetary space. As I have mentioned earlier, some magnetic field lines in the corona return rapidly to the surface of the Sun, while those passing through coronal holes go out into interplanetary space. It is found that the solar wind flows out along the open field lines, as must be an inevitable consequence of the motion of charged particles being tied to magnetic field lines. The overall conclusion is that the corona is hotter along the closed field lines than along the open field lines and that the input of energy from below is divided differently between coronal heating and driving mass loss through the solar wind on different field lines. In Chapter 6, we shall see that there is

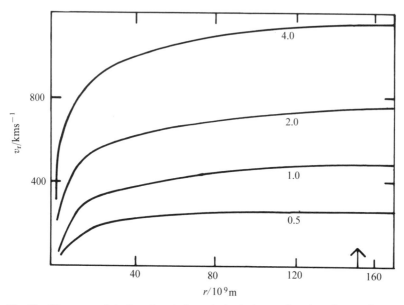

Fig. 29. The approach to the solar wind terminal velocity as a function of coronal temperature (based on E N Parker's original calculations). The curves are marked with the coronal temperature in units of 10^6 K. The arrow on the horizontal axis shows the Earth's distance from the Sun.

some division between stars which have high mass loss rates as red giants and those which have hot coronae.

The solar wind only exists because the Sun has a hot corona. It is now necessary to ask why the corona is so hot.

The heating of the corona

The original idea for the heating of the corona was entirely non-magnetic and it was roughly as follows. It is known from laboratory experiments that, if a fluid is set into violent motion, it emits sound with the amount of sound rising as a high power of the average velocity of the fluid. The outer layer of the Sun is observed to be in a state of convective motions. If these convective motions produce sound waves, these should propagate outwards from the surface of the Sun. A wave motion does not involve any net flow of material outwards, but the wave motions have an energy density, $\frac{1}{2}\rho v^2$, where ρ is the density of the matter and v^2 is the mean square velocity of the particles executing the wave. This energy is conserved as the wave moves outwards and, if the wave moves into a region of much lower density, the wave must become much more intense. The increase in wave amplitude produces very steep gradients in the values of the physical quantities involved. The wave eventually turns into a shock wave and there is a strong dissipation of energy, which is converted into heat and which raises the local temperature. This acoustic heating of the corona was a popular idea for some years but it was eventually accepted that it was not efficient enough to produce the observed hot corona. Acoustic heating may be important in the outer layers of some other stars. Now it is generally accepted that magnetic fields play a key rôle, although there is still no clear agreement about what is the precise mechanism.

There are two main competitors. The first is also wave heating but heating produced by magnetohydrodynamic waves rather than sound waves. The second involves dissipation of energy by electric currents possibly produced by reconnection of magnetic fields, as discussed in Chapter 4. As has been made clear in Chapter 4, when a magnetic field is present, there are two characteristic speeds of wave propagation, the sound speed, c_s, and the Alfvén speed, c_H. Only when there is no magnetic field present can pure sound waves propagate, except in the direction of the magnetic field. Even the latter ceases to be exactly true when the field direction is not constant. If $c_s \gg c_H$ magnetic effects are a small perturbation, but that is not the case in the outer solar atmosphere where it is much more likely that $c_H > c_s$. This implies that it probably does not make sense to discuss a purely acoustic mechanism for heating the corona.

If the corona is to be heated by magnetohydrodynamic waves, the basic idea is very much the same as that of acoustic heating. Disturbances in the photosphere must lead to wave motions propagating into the upper atmosphere. The waves must then steepen and dissipate. There are detailed differences between the acoustic and magnetic processes. In particular, as discussed in Chapter 4, magnetohydrodynamic waves have an essentially anisotropic propagation, whereas sound waves would spread out spherically if it were not for spatial variations in the properties of the medium in which they propagate. Magnetohydrodynamic wave heating is regarded as a good candidate for a coronal heating mechanism but it has not been proved to be responsible .

Another favoured mechanism is magnetic reconnection. As I have already explained many magnetic field lines leave the solar photosphere, rise up in the chromosphere and corona and then return to the photosphere again. The configuration of the magnetic field is far from static. If I call the points where the field lines enter and leave the photosphere the footpoints of the field lines, these footpoints are being continually moved around by the convective motions in the solar atmosphere. The field line arches above the photosphere must respond to the motions of the footpoints and it is possible that two field lines are brought to a position where they are trying to cross one another (fig. 30). In this situation, magnetic reconnection of the type which I described in the last chapter can occur and magnetic energy is dissipated as heat. In the reconnection process electrical currents flow. There have been other proposed coronal heating mechanisms, which rely on the production of electrical currents in a different way.

Reconnection may be a more discrete and discontinuous process than wave heating and ultimately a decision as to which process is relevant (unless they both are) could come by an observation of individual reconnection events in the right place. The heating mechanisms must not only be capable of releasing the correct total unit of energy but

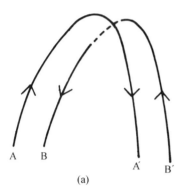 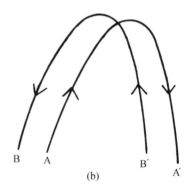

Fig. 30. In (a) two field lines with footpoints AA′, BB′ are such that one is behind the other. In (b) convective motions have moved the footpoints causing the field lines to intersect so that reconnection can occur.

they must provide it where it is wanted. Thus they must produce the hottest corona in the region of the closed magnetic loops rather than on the open field lines.

A recent space mission has provided some very unexpected new information about the solar wind. All previous observations have been of the solar wind in or near to the ecliptic plane which contains the Sun and the planets. The *Ulysses* mission first travelled to Jupiter in the ecliptic and then received a gravitational impulse which sent it on an orbit far out of the ecliptic plane, so that it passed below the South pole of the Sun in 1994 and above the North pole in 1995. Amongst its discoveries were that at solar minimum, which was when it passed the poles, they were both covered with coronal holes. From these coronal holes was flowing a solar wind with a significantly higher speed than that in the ecliptic. The typical speed was found to be about 700 km s^{-1} rather than 400 to 500 km s^{-1} in the ecliptic. This is yet a further demonstration of the anisotropic nature of the wind.

Before I leave the solar wind and the corona, I should make a general comment on the source of the energy that they carry away from the Sun, in coronal radiation and in the kinetic energy carried away by the solar wind. In total the corona and the solar wind (and indeed the flares which I shall discuss shortly) radiate an amount of energy which is very small compared to the photospheric luminosity of the Sun. However, they all have the same ultimate origin, which is the nuclear reactions occurring in the solar core. In the simplest non-magnetic model of the Sun, about 10^7 years supply of luminosity is stored in the thermal energy of the Sun. In a magnetic model the dynamo process causes some thermal energy to be converted into and stored as magnetic energy. Both the thermal energy and the magnetic energy then contribute to the total luminosity of the Sun, when they are converted into radiated energy.

The chromosphere and transition region

Between the photosphere and the corona lie the *chromosphere* and the *transition region*. Immediately above the photosphere the temperature drops to the temperature minimum and then it rises again first slowly in the chromosphere and then very rapidly through the transition region. This has been illustrated in fig. 8. In all these zones the temperature is not a true thermodynamic temperature but it is a kinetic temperature which is a measure of the kinetic energy of the particles. The chromosphere was so-named in 1869 after it had been seen in observations of a total solar eclipse. For just a few seconds after the onset of totality a vivid red streak was observed close to where the photosphere had last been visible. Similarly the red streak was observed just before totality ended. It was named the chromosphere because it was so colourful. The chromosphere is about 10 000 km thick or about $1\frac{1}{2}$ per cent of the radius of the Sun.

The early detailed observations of the chromosphere related to the low chromosphere where the kinetic temperature is not much higher than the photospheric temperature. The photographs were frequently taken in the light of Hα, the lowest line in the Balmer series of hydrogen. Hα appears as an emission line in the chromosphere and in an Hα photograph the sunspots appear dark and they are surrounded by bright regions known as *plages*. There are also elongated dark elements known as *mottles* and *fibrils*. It was seen in early photographs of the chromosphere taken with ordinary photographic plates, that the chromosphere is not a uniform flat layer but that it contains extended elements known as *spicules* which wave about like grass in a wind. In the lower part of the chromosphere, as in the photosphere, the magnetic field is only important in limited regions. As the gas density drops rapidly with height in the outer solar atmosphere, while magnetic flux is conserved, magnetic flux tubes must expand in order to maintain pressure balance in the atmosphere. Eventually the magnetic field must become essentially *force-free*. This means the electric currents have the same direction as the magnetic field so that the force curl $\mathbf{B} \times \mathbf{B}$ is zero. No significant deviation from this property is possible because the gas is not dense enough to produce a pressure gradient which can balance any significant magnetic force. In the region where the field is effectively force-free, the magnetic field fills most of the atmosphere as it does in the corona.

The other very important emission lines observed in the chromosphere are the so-called H and K lines of singly ionised calcium. The properties of these lines are very strongly correlated with the magnetic field structure in the chromosphere. As a result they provide very good evidence for magnetic activity. As I shall explain in the next chapter, their discovery in other stars was one of the first clear pieces of evidence for stellar magnetic activity and the variation of their strength with time suggests the existence of magnetic cycles similar to the solar cycle.

The high chromosphere has a kinetic temperature measured in tens of thousands of degrees and then in the transition region there is a very rapid transition from about 5×10^4 K to a few million K. Obviously if the corona has a temperature of millions of degrees, there must be a continuous transition between that temperature and the photospheric temperature of close to 6000 K. What is not immediately clear is why the transition has the form that it does and what is the heating mechanism of the chromosphere and the transition region. More specifically, is the heating from below or is it from above? There is not at present a clear answer to this question although it is quite possible that both wave heating from below and conduction from above play a part.

Solar flares

In Chapter 2, I have given a brief account of the properties of solar flares and I have suggested that their energy release involves the reconnection of magnetic field lines which I have discussed in Chapter 4. I shall now say a little more about current ideas concerning the origin and energies of solar flares. A first comment is that developments in the spectral range of observation of the Sun have led to a change in the perceived location of a flare outburst. When flares were first discovered in the integrated solar emission (*white light flares*) and then in the Hα emission line of neutral hydrogen, they were thought to be a chromospheric phenomenon; indeed they used to be known as chromospheric flares. The developments of observations in X-rays and the ultraviolet have shown that the main energy release in flares actually occurs in the corona and that the observed chromospheric activity arises by transfer of energy from the corona to the chromosphere.

Flares are far too complex a phenomenon for me to attempt to discuss either the detailed observations or the theoretical interpretation in a book at this level. I shall therefore give a qualitative description of what is observed and of what is believed to be behind the observations. An important component of flares is energetic particles. If, as I have suggested, a flare is associated with a rapid time variation of a magnetic field, there must be an associated electric field and it is therefore not surprising that charged particles can be accelerated to high energies. Direct observations of high energy solar flare particles can be made at Earth as will be mentioned in Chapter 7. This solar component of the low energy cosmic rays can be very important at the time of a very intense solar flare. At a time when only the low energy cosmic rays had been well-studied, a minority of astronomers believed that the Sun and similar stars might be the main source of cosmic rays. This is no longer thought possible but they do provide a contribution at low energies.

The high energy charged particles have a variety of other observed effects. The energetic protons can collide with atomic nuclei and either knock particles out of them or raise the nuclei to excited states. The first process can produce energetic neutrons. The second process can lead to some of the observed γ-ray emission from flares. For example ^{12}C and ^{16}O have energy levels 4.4 MeV and 6.1 MeV above their ground states and γ-ray lines observed at these energies can be attributed to emission when excited nuclei return to their ground states. Positrons can also be produced by various reactions and they subsequently annihilate with electrons to produce a γ-ray line at 0.511 MeV; the rest mass energy of the electron and the positron is converted into two photons with this energy. There are several types of radio emission associated with flares and this radiation can be associated with energetic electrons moving in a magnetic field. Finally energetic particles moving down the magnetic field lines can carry the flare energy down to the lower atmosphere and produce the chromospheric emission which was the original signature of the flare. A white light flare occurs if the heating extends as far down as the photosphere.

X-rays are very prominent in flares and these indicate that an intense flare produces a region of hot plasma with a temperature of several times 10^7 K, which is more than ten times hotter than the normal corona. It is possible that a small number of particles may acquire sufficient energy to collide and undergo nuclear fusion reactions, but this is not a major factor. There may also be some breakup or spallation reactions which I shall mention in connection with stellar flares in Chapter 6. In a significant fraction of flares, a rising filament of material is observed in the corona and this can lead to loss of mass from the Sun in what is known as a coronal mass ejection. The total loss of mass in these discrete explosive events is comparable with the loss of mass in the solar wind and this is another reason why the description of the solar wind, which I gave earlier in this chapter, is oversimplified.

Multiwavelength observations with good spatial and temporal resolution can be made of solar flares particularly when satellites, which work in regions of the spectrum which are not accessible from the Earth's surface, are operational at the same time. This means that the development of a flare can now be very well documented and this provides a formidable amount of information for the theoretician to attempt to interpret. Thus, for example, it is possible to compare the emission in both γ-rays and X-rays from a flare. Their general behaviour is similar but the peaks of γ-ray emission tend to occur shortly after peaks in hard X-ray emission. Radio emission also starts one or more minutes after the onset of the γ-rays and X-rays. The brightening of the chromosphere as energy is carried down to it then provides information about the manner in which this occurs.

Flare energy release

Now I turn to the source of the flare energy and to the manner in which it is released. The timescale of a typical flare ($\sim \frac{1}{2}$ hr) is short compared to the time in which significant effects occur in the solar photosphere. This means that the footpoints of the magnetic field lines, which penetrate into the corona where the energy release occurs, must be in effect frozen into the photosphere. This means that any rearrangement of magnetic fields with the release of energy must have as a boundary condition the constraint of no change in the field emerging from the photosphere. Not all of the energy stored in the magnetic field loops can be released. The most that can possibly happen is that the field relaxes to a current-free configuration consistent with the prescribed photospheric field.

I have said several times that the liberation of energy probably takes place as a reconnection of magnetic field lines. The description of reconnection which I have given in Chapter 4 is highly simplified and it does not obviously satisfy the requirements for solar flares. If field lines are being brought together continuously one might expect to have a steady release of energy over a rather long timescale rather than the explosive release which occurs in an intense flare. We shall see in Chapter 7 that such steady reconnection is probably relevant in the case of the interaction of the solar wind with the Earth's magnetosphere. For a flare we require the magnetic field structure to evolve without any significant energy release until something dramatic occurs.

One possibility is that the magnetic field configuration evolves into a state in which it is unstable and that this instability evolves on a very short timescale and leads to magnetic field reconnection with release of energy. It is very well known that configurations of plasmas and magnetic fields are frequently unstable. This was discovered in the 1950s when the first serious investigations were made of the possibility of producing controlled thermonuclear reactions in the laboratory by confining a hot plasma with a magnetic field. It was soon found both experimentally and theoretically that the proposed configurations were unstable. Typically the instability growth times were comparable with l/c_H, where l is the characteristic length of the apparatus and c_H is the hydromagnetic velocity. These instabilities occur in a plasma in which electrical resistivity is unimportant. If resistivity is important, there are new resistive instabilities which have growth times of order $(\tau_H \tau_R)^{1/2}$, where τ_H is the growth time given above and τ_R is a resistive diffusion time. One particular resistive instability known as the tearing mode is a candidate for the trigger that leads to a sudden release of energy. A discussion of a variety of models for reconnection can be found in the book *Solar Magnetohydrodynamics* by E R Priest which is listed in the Suggestions for Further Reading at the end of the book.

I mentioned earlier that reconnection may also be the source of heating of the solar corona. In that case what is required is a much less dramatic release of energy and one that is distributed much more widely over the solar surface. Coronal heating, if it is not due to the dissipation of magnetohydrodynamic waves, might result from the occurrence of a very large number of microflares, very low level reconnection events which would not qualify for the status of normal solar flares.

I now turn to a suggestion as to how rising filaments and coronal mass ejections might be produced. Suppose that there is a collection of intertwined flux loops which are connected to the photosphere, as shown in fig. 31a. If movement of the footpoints leads to oppositely directed field lines being brought together, reconnection can occur which reduces the number of footpoints of the now combined loop or filament (fig. 31b). The large release of energy associated with the reconnection can then cause the loop to rise and in some circumstances this might become a coronal mass ejection.

The theoretical description of the explosion of a solar flare is very complex and beyond the level of the present book but I will complete this section by showing a diagram of a model of what is known as a two-ribbon flare taken from a paper published in 1990 (fig. 32). This diagram shows phenomena occurring all the way from the corona down to the photosphere. The main energy release is occurring in the current sheet in the mid corona and from that level influences are propagating both inwards and outwards. There is a white light flare visible at the photosphere and a coronal transient in the high

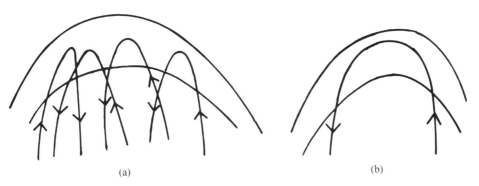

(a)

(b)

Fig. 31. In (a) a filament is attached to the photosphere by a complex of flux loops. In (b) a succession of reconnection events has left the filaments connected by a single loop and this makes its upward motion easier.

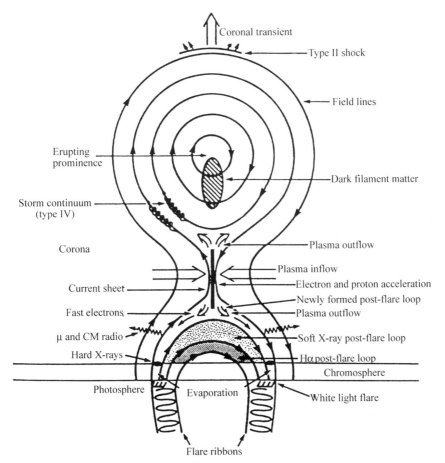

Fig. 32. A detailed solar flare model from Martens, P C H & Kuin, N P M, *Solar Physics*, **122**, 263, 1989. The main energy release occurs through reconnection in the current sheet in the middle of the diagram. From this region both energy and matter flow both upwards and downwards as shown.

corona. A major problem in proposing and checking such a flare model is that observations in any particular spectral range only provide information about a cross-section through the flare and multifrequency observations are required to provide all of the information about a flare that has to be checked against the theoretical model.

The production of high energy particles in solar flares has important consequences on Earth as does the solar radio emission. The magnetic storms associated with enhanced solar wind produced during flares will be discussed in Chapter 7, as will the enhanced auroral emission, which is probably caused by the more energetic solar particles. The intense radio emission from the Sun was discovered during the Second World War. In February 1942 the British radar system was put out of action by what first appeared to be deliberate *jamming* by the Germans and indeed several German battleships were as a result able to slip through the English Channel without detection. In fact it was soon discovered that the jamming had a natural origin. The intense radio emission came from the Sun and it was quite by chance that the German ships gained an advantage from it. This was one of a number of wartime discoveries, which led to the development of radio astronomy after the war.

Prominences

One of the most dramatic events observable on the Sun during an eclipse is a giant prominence (fig. 33). These appear as large bright arched regions which expand

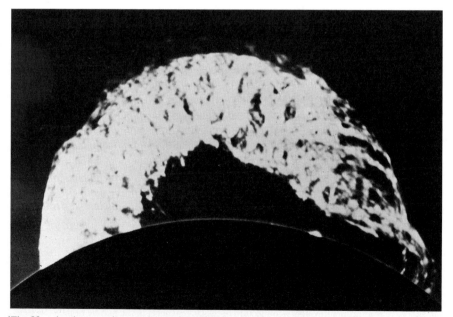

Fig. 33.   A solar prominence observed on 1946 June 4. (Royal Astronomical Society.)

outwards from the solar surface. Obviously, for a prominence to be observed in this manner, it must be at the limb of the Sun at the time of the eclipse. Because of the solar rotation, there is a relatively small chance of this being the case, so that it is obvious that during an eclipse, or in observations of an artificial eclipse using a coronagraph, only a small minority of the solar prominences can be observed. In fact this is not quite true. When the Sun passes out of the eclipse, the solar rotation can cause some prominences to lie between the observer and the photosphere. In this case, as has already been mentioned in Chapter 2, the prominences are observed as dark streaks across the visible surface.

This observation implies that the bright prominences must be cooler than the photosphere and even more so than the chromosphere and the corona into which they are penetrating. They can appear as bright as they do in eclipse observations both because they must be much denser than the hotter gas in which they are embedded and because this hotter gas has its principal emission in a different region of the spectrum. The shape of a rising prominence strongly suggests that it is threaded along a filament of magnetic field. That leaves two questions to be answered. How is a prominence produced and why does the prominence material not fall back to the solar surface, given that it is heavier than its surroundings? The largest prominences extend a significant fraction of a solar radius above the photosphere.

Although the most dramatic prominences are those which are known as *eruptive prominences*, which are observed to rise rapidly upwards from the solar surface, there is another class known as *quiescent prominences* which have a much longer lifetime which is typically several solar rotation periods and it is easier to discuss their properties. A quiescent prominence is a hugh arch of dense cool material embedded in the hot corona. The length of the arch is typically several hundred thousand kilometres, which compares with a solar radius of 7×10^5 km. The height is up to 10^5 km; if a quiescent prominence changes into an eruptive prominence, the height can become much greater. A typical thickness of the loop is 10^4 km. At the end of its life a prominence disperses and breaks up quietly, or it becomes an eruptive prominence or matter falls back down the field lines to the photosphere.

A mechanism for the formation of cool prominence material in the corona is *thermal instability*. The basic idea of thermal instability is quite easily explained, although the details are very complicated. The equilibrium of the solar corona requires the heating and cooling processes to be in balance. Earlier in this chapter I discussed ideas relating to the heating of the solar corona. What I did not discuss was precisely what determines the coronal temperature. At any given time the heat input to the corona must be balanced by the energy radiated by the corona, which escapes from the Sun, and by heat conduction from the corona to the layers below. Suppose now the equilibrium of the corona is disturbed so that locally the density is increased. It is likely that the cooling processes will increase as the density increases and as a result the denser region will become cooler than its surroundings. Provided thermal conduction from the hotter surroundings cannot restore the equality of temperature, the dense region will continue to cool until it reaches a new equilibrium in which its heat input balances its heat output. In this manner thermal instability has produced a dense cool sheet of gas in the corona.

Two important factors have not been included in the above qualitative discussion, gravity and the magnetic field. The magnetic field is obviously important because of the influence which it has on the motion of charged particles. Thermal conduction results from electrons moving from hot regions to cooler regions and exchanging energy through collisions. In the presence of a magnetic field, the particles can move freely along the field but only with difficulty across the field. This implies that the thermal conductivity, $\kappa_{\parallel}$, along the magnetic field is very much greater than the conductivity, $\kappa_{\perp}$, across the field. As a result the longest dimension of any cool material is likely to be along the field.

Prominence support

The first prominence model to include gravity was produced by R Kippenhahn and A Schlüter in 1957. This model was simplified in another manner in that it essentially ignored the thermal properties of the prominence. Consider first the equation of equilibrium of a magnetised fluid acted on by gravitational field, g, in the z direction. The equation of equilibrium is

$$0 = -\mathrm{grad}\, P - \rho g \hat{\mathbf{z}} - \mathrm{grad}(B^2/2\mu_0) + \mathbf{B}.\nabla\mathbf{B}/\mu_0, \tag{5.13}$$

where P and ρ are the pressure and density of the fluid, $\hat{\mathbf{z}}$ is a unit vector in the z direction and the magnetic force is taken from equation (4.17). This must as usual be supplemented by the equation

$$P = \mathscr{R}\rho T/\mu, \tag{5.14}$$

where T is the temperature and μ is the mean molecular weight of the material. In their first simple model Kippenhahn and Schlüter assumed that the temperature T and the horizontal magnetic field components, B_x, B_y were constant and that P, ρ, and B_z were functions of x alone. Thus the prominence is represented as a plane sheet, which also is not limited in the direction of the gravitational field. This can obviously only provide a first crude idea of the mechanism of prominence support.

The only equations which are not satisfied identically as a result of these assumptions are the x and z components of equation (5.13). These are

$$0 = -\frac{\mathrm{d}}{\mathrm{d}x}(P + B^2/2\mu_0). \tag{5.15}$$

$$0 = -\rho g + (B_x/\mu_0)\mathrm{d}B_z/\mathrm{d}x. \tag{5.16}$$

These equations have to be solved with the boundary conditions

$$P \to 0, \quad B_z \to \pm B_{z\infty}, \quad \text{as } x \to \pm\infty. \tag{5.17}$$

Equation (5.15) implies that the total pressure, gas plus magnetic, is constant. Combined with (5.17), this gives

$$P = (B_{z\infty}^2 - B_z^2)/2\mu_0. \tag{5.18}$$

Equation (5.14) then gives an expression for ρ. If this is inserted into (5.16), there results

$$0 = -(B_{z\infty}^2 - B_z^2)/2\Lambda + B_x dB_z/dx, \tag{5.19}$$

where

$$\Lambda = \mathcal{R}\rho T/\mu g. \tag{5.20}$$

Equation (5.19) is then readily solved to give

$$B_z = B_{z\infty} \tanh(B_{z\infty}x/B_x\Lambda), \tag{5.21}$$

while equation (5.18) yields

$$P = (B_{z\infty}^2/2\mu_0) \operatorname{sech}^2(B_{z\infty}x/B_x\Lambda). \tag{5.22}$$

The horizontal structure of the pressure and the z component of the magnetic field is shown in fig. 34. It is clear that the y component of the field is completely passive and

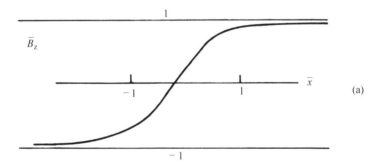

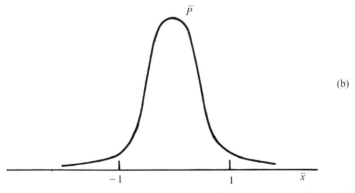

Fig. 34. The dependence of magnetic field and plasma pressure on distance from the prominence centre in a simple model by Kippenhahn, R & Schlüter, A.

plays no rôle in the solution. In contrast the x component determines the thickness of the sheet, assuming that all the other parameters are fixed.

A realistic prominence model must contain many factors which are not included in the above very simplified model. Clearly the first thing to include is an energy equation which allows for the large difference in temperature between the prominence and its surroundings. The simplest approach is to add an energy equation of the form

$$\frac{d}{dx}\left[\kappa_0 T^{5/2}\frac{dT}{dx}\frac{B_x^2}{B^2}\right] = C - H, \tag{5.23}$$

where the expression for thermal conductivity used earlier in the chapter in the discussion of the solar wind has been modified by the factor B_x^2/B^2 because conduction must follow the field lines. C is the radiative cooling term and H is the heating term. Expressions for C and H in terms of ρ, T and the chemical composition of the corona can be inserted. In this approach it is assumed that all radiation escapes freely to infinity. A further subtlety involves solving for the transfer of the radiation including absorption and scattering.

A discussion of the solution of these equations is outside the scope of this book. However, once the more realistic prominence models have been produced, it is necessary to test them for stability. It is no use demonstrating that an equilibrium is possible, if the equilibrium proves to be unstable on a timescale which is very short compared with the observed prominence lifetimes. One possibility is that a prominence magnetic field configuration is initially stable, allowing for the existence of a long-lived quiescent prominence but that the field evolves into an unstable configuration which is followed by prominence collapse or eruption. Further discussion of prominence models can be found in the more advanced books listed in the Suggestions for Further Reading at the end of this book.

Sunspots

In Chapter 2 and earlier in this chapter I have given a brief description of the observed properties of sunspots and of their relation to the solar dynamo. In the remainder of this chapter, I now say something about the structure of individual sunspots. The ideal theoretician's sunspot is an isolated tube of magnetic flux which penetrates the surface of the Sun (fig. 35). The visible sunspot is situated in a depression in the photosphere and it is cooler than its surroundings. Reality is more complicated. Sunspots typically occur in pairs or more complicated groupings and individual sunspots show strong departures from symmetry. The simple model is however useful in getting ideas straight.

It is now generally accepted that the darkening of a sunspot is produced by the interference with convection by a magnetic field, as was first discussed by L Biermann in 1941. It is easy to see that convection must be modified if not suppressed by a field. The cellular motion which is observed in the normal granulation cannot occur if material is required to cross magnetic field lines. If the motion is strong enough it may distort the

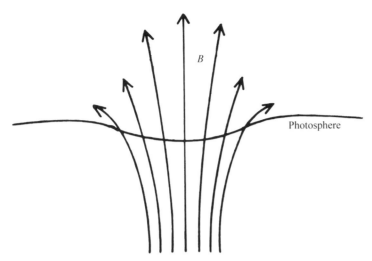

Fig. 35. A theoretician's idealised sunspot model. The Wilson depression in the photosphere is shown. The magnetic flux tube is axisymmetric. The magnetic field is vertical in the middle of the umbra and moves towards the horizontal in the penumbra.

magnetic field but if it is weaker it will be suppressed by the field. As I have discussed in Chapter 4, the condition for the onset of convection is modified in the presence of a field and the observed magnitude of sunspot fields is such that suppression of convection would be expected. The absence of normal granulation inside the sunspots is supporting evidence. For a time there was worry about this model, because it was thought that the energy, which is not carried upwards by convection in a spot, should appear as a bright ring immediately surrounding the spot. Such a bright ring is not observed. There is now general agreement that the energy diffuses sideways in such an efficient manner that it does not make any noticable change in the appearance of the photosphere. In fact it is now believed that this energy is stored in the convection zone for a time which is much longer than the lifetime of an individual sunspot, so that there is no way in which the consequences of convective flux blocking in the spot can be studied. Although motions across the magnetic field are suppressed, there remains the possibility that there will be some oscillatory motions along the field lines. These will not be anything like so efficient at carrying heat as normal convection.

Sunspot structure

Only the surface properties of the flux tube that defines a spot can be studied and how the field structure changes with depth can only be deduced from theoretical considerations. The simplest model is a monolithic column of flux (fig. 36) and I will start with a discussion of such a structure. The model I describe is one in which it is assumed that the pressure inside the flux tube is negligible compared to the magnetic pressure. This is an

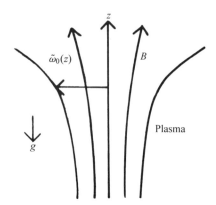

Fig. 36. A model magnetic flux tube embedded in field-free plasma (based on calculations by Meyer, F, Schmidt, H U & Weiss, N O, *Monthly Notices Royal Astronomical Society,* **179**, 741, 1977).

extreme assumption but may provide a qualitatively correct picture. If that is the case and if it is also assumed that the gravitational force is unimportant in obtaining an approximate idea of the magnetic field structure, the magnetic field in cylindrical polar coordinates can be taken to be current free, so that

$$\mathbf{B} = \frac{1}{\tilde{\omega}}\left[-\frac{\partial \psi}{\partial z}, 0, \frac{\partial \psi}{\partial \tilde{\omega}}\right]. \tag{5.24}$$

It is easily verified that curl $\mathbf{B} = \mathbf{0}$. In addition div $\mathbf{B} = 0$ implies that

$$\frac{\partial^2 \psi}{\partial \tilde{\omega}} - \frac{1}{\tilde{\omega}}\frac{\partial \psi}{\partial \tilde{\omega}} + \frac{\partial^2 \psi}{\partial z^2} = 0. \tag{5.25}$$

In principle the neighbouring photosphere, in which the flux tube is embedded, has a known pressure variation with height, $P_e(z)$ and the boundary of the flux tube should be placed at $\tilde{\omega} = \tilde{\omega}_0(z)$, where

$$B^2/2\mu_0 = P_e(z). \tag{5.26}$$

As $z \to \infty$, the field becomes nearly horizontal with $B_{\tilde{\omega}} \approx F/2\pi\tilde{\omega}_0^2$, and as $z \to -\infty$, the field becomes vertical with $B_{\tilde{\omega}} \approx F/\pi\tilde{\omega}_0^2$; in each case these values are those on the boundary. F is the total flux which is $2\pi\psi$ evaluated on the boundary. The easiest approach to a solution of the problem is to prescribe a solution to equation (5.25) and then to determine $P_e(z)$ and $\tilde{\omega}_0(z)$. This approach provides model flux tubes similar to that shown in fig. 36 but they do not give a very good fit to the run of $P_e(z)$ in the solar atmosphere. The problem becomes much more complicated, if both the internal pressure and the gravitational force are introduced, but this is necessary for a fully realistic model. The internal pressure must obviously be included if there is to be a realistic model for the Wilson depression in the spot.

One problem with the simple monolithic structure of a flux tube is that, although the spots are cooler than the surrounding photosphere, the difference in the energy radiated by the spot and by an equivalent area of the normal photosphere is only about a factor of four; it looks more extreme in the visible because the lower spot temperature means that a smaller fraction of its radiation is in the visible. This factor of four is less than would be

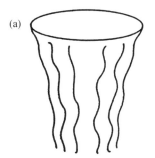 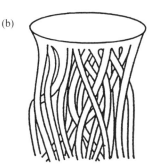

Fig. 37. Two possible magnetic field behaviours below a sunspot. (a) coherent flux tube and (b) tight cluster (diagram courtesy of Thomas, J H and Weiss, N O).

expected if convection in the spot were completely suppressed throughout the depth of a single flux tube. I have not at present tried to define where the bottom of the spot should be placed but a first guess would be the level at which the gas pressures inside and outside the flux tube are approximately equal. At such a level, the magnetic field can no longer control the dynamics. This depth is in any case well below the visible photosphere. If convection were suppressed throughout the depth of the spot, radiation would not carry enough energy to provide the observed temperature of the umbra.

It is therefore believed that some form of convective energy transport must occur not far below the visible surface. As such energy transport is not compatible with the regular monolithic magnetic field structure, it is necessary to suppose that the field is more complicated. There are several suggestions about how the field is arranged. In the first model the flux tube retains its entity but the field lines are distorted as a result of fluid motion. In a second model, it breaks up into a tight bundle of isolated tubes which have field-free plasma between them. These models are sketched in fig. 37.

Sunspot oscillations

The sunspot umbra at low resolution appears to be more or less uniformly dark but, when the resolution is increased, this ceases to be true. An umbra can then be observed to contain a large number of bright points known as umbral dots. It is believed that these umbral dots provide evidence for the oscillatory motions up and down the field lines, whose possible existence has been mentioned earlier. Theoretical models of fully developed convection in the presence of a magnetic field (magnetoconvection) provide important clues to the complex structure of sunspot umbrae and penumbrae, although there is not yet a full understanding of what is occurring.

It is possible to observe a variety of coherent oscillations associated with sunspots. In the first instance the five-minute solar oscillations which were discussed in Chapter 3 can be observed in sunspots. The oscillations which are observed for the whole Sun have largely travelled through regions in which the magnetic field provides a small perturbation to the non-magnetic oscillations. In contrast, if the oscillations are observed at the photosphere of a sunspot umbra, they have recently passed through a region in which the magnetic field is very important. As waves which can propagate in a sunspot must basically be magnetohydrodynamic waves of the type which I have discussed in Appendix 6, they should behave differently from waves in the surrounding photosphere.

This is what is observed. The observed power of the five-minute oscillations in both sunspots and other active regions, where the magnetic field is strong, is significantly less than what is observed over other parts of the solar surface. Although there is no complete agreement about precisely what is the cause of this reduction, it appears very reasonable that a sunspot or other region with a strong magnetic field should provide a filter to the normal solar oscillations and only allow the transmission of those waves which match appropriately to the magnetic properties.

A sunspot is usually thought of as a photospheric phenomenon, because that is where it was originally observed in visual observations. It is however obvious from what I have said earlier that the flux tube must penetrate up into the chromosphere. In the chromosphere associated with spots, other important oscillations have been observed. These oscillations can be observed by studying the behaviour of the spectral lines of ionised calcium in the chromosphere. The central wavelength of the Ca K spectral line is observed to vary with time and this variation can be interpreted as a Doppler shift due to wave motion. The shift is asymmetric with a rapid shift in the direction from red to blue with a more gradual return. This is evidence for a non-linear wave motion rather than a low amplitude approximately sinusoidal wave. These oscillations usually have a frequency of about 5.5 mHz and there is no sign of the photospheric five-minute oscillations in the chromosphere. The origin of these oscillations is not fully understood although several theoretical explanations have been proposed. They potentially provide very valuable information about the formation of large amplitude waves in the presence of a magnetic field.

Sunspot stability

Sunspots live much longer than the typical dynamic time associated with their structure; this is approximately the time taken by a magnetohydrodynamic wave crossing the spot. This means that sunspots must be stable structures, at least with regard to the potentially worst type of instabilities. There are two competing factors in the stability of simple spot models of the type which I have described above. If gravity were absent the spot configurations would be unstable. It is well known, from studies of stability concerned with potential controlled thermonuclear reactors, that instability occurs if a dense plasma has a surface which is convex towards a magnetic field. That is the case in configurations like those of fig. 36. The instability is known as an interchange instability because the effect is for plasma to interpenetrate the field lines which themselves move outwards. The second factor is stabilising. What is known as a Rayleigh–Taylor instability occurs if a heavy fluid sits on top of light fluid in a gravitational field. The fluids change position. In the case of the sunspot, its interior forms the light fluid and at the surface of the flux tube the light fluid is on top, which is the stable position.

In a real sunspot model a careful calculation needs to be made. There is a general mechanism for studying stability of static magnetohydrodynamic equilibrium. If it is supposed that the equilibrium is perturbed so that an element of material originally at vector position $\mathbf{r}_0$ moves to

$$\mathbf{r} = \mathbf{r}_0 + \boldsymbol{\xi}(\mathbf{r}_0), \tag{5.27}$$

it is possible to write down an expression for the change in potential energy of the system produced by the perturbation. Because the system is initially in equilibrium, the part which is linear in $\boldsymbol{\xi}$ vanishes identically. For stability it is then necessary that the part that is quadratic in $\boldsymbol{\xi}$, δW, should be positive for all possible perturbations $\boldsymbol{\xi}$. Instability is demonstrated, if there is any $\boldsymbol{\xi}$ for which $\delta W < 0$.

It is obviously easier to demonstrate that an unstable system is unstable than to prove that a stable system is stable. δW is an integral over the entire relevant volume of a quadratic fuction of $\boldsymbol{\xi}$ and its derivatives and involving $\mathbf{B}$, P, ρ and g. The stability of the more simple spot models has been studied using this technique. In general it appears that the models are stable provided that they contain a large enough magnetic flux. Slender flux tubes appear to be unstable and this suggests that the observed distribution of size of sunspots may be influenced by stability considerations.

Sunspot decay

It has been realised for a long time that, if sunspots are basically stable objects, their eventual decay cannot be due to the ohmic decay of the sunspot magnetic field. The calculated magnetic diffusion time (equation 4.13) for a sunspot magnetic field is about 10^3 yr compared with sunspot lifetimes of a fraction of a year. The normal static decay must be enhanced by some hydrodynamic process. What is required is that some motions, which might be driven by a late developing instability, produce a turbulent resistivity of the type discussed in dynamo models (equation 4.49) and that this reduces the decay time. In some models the motions start in the penumbra and move inwards. Another process which is observed is that many sunspots *fragment* rather than decay. They break up into several parts and these smaller flux tubes are able to decay more readily than the single large tube.

My discussion of the active Sun has necessarily been very sketchy. It is to be hoped that it is adequate to provide an introduction to treatments which try to take account of the full complex observations.

Summary of Chapter 5

The surface activity of the Sun is driven by the solar magnetic field, which is itself generated by dynamo action in or just below the solar outer convection zone. Direct evidence about the properties of the dynamo is provided by the behaviour of the sunspot cycle with its approximate 22 year periodicity. The Maunder minimum of sunspot activity in the 17th century and indirect evidence for earlier minima, suggest that the solar dynamo is chaotic. Although the sunspots are the most obvious evidence for the dynamo, most of the magnetic flux which penetrates the solar surface is not concentrated in sunspots. At present theoretical models of the solar dynamo are not capable of a full explanation of the observed properties of the solar cycle. This means that, in addition, the structure of the magnetic field and convection throughout the outer convection zone is not well-understood.

The Sun loses mass continuously in the solar wind. Its existence is a consequence of the hot corona. It is not possible for a static hot corona to extend throughout interplanetary space. It must expand and as a result the Sun loses mass. The solar wind is far from being spherically symmetric. The corona is composed of closed magnetic loops and of open field lines along which the solar wind flows. It is necessary to understand why the corona is so hot. It is believed that the energy input must be provided by the magnetic field either through the dissipation of magnetohydrodynamic waves or through the reconnection of magnetic field lines. Between the corona and the photosphere lie the chromosphere and the transition region. The chromospheric magnetic activity is apparent in emission lines of hydrogen and calcium and the observation of similar emission lines in other stars was one of the earliest evidences for their magnetic activity.

Solar flares represent a more violent and irregular form of magnetic activity. Although flares were originally observed in the chromosphere, it is now clear that they range all the way from the corona to the photosphere. Only the liberation of magnetic energy can provide their radiation and the process involved is magnetic field reconnection. Flares are evident in all regions of the electromagnetic spectrum from the γ-ray to the radio. The radio emission is produced by energetic charged particles and some particles accelerated in flares are subsequently observed at Earth. Flares are sometimes associated with a loss of mass from the Sun known as a coronal mass ejection and these ejections provide a mass loss from the Sun compared with that of the steady solar wind.

Another dramatic event observed on the Sun is a giant prominence. This is a large arc of cool material threaded along a magnetic field loop and embedded in the very much hotter corona. Some prominences are long lived and quiescent whereas others erupt and rise rapidly. There are problems associated with the formation, equilibrium and stability of prominences. Quiescent prominences may be formed by a thermal instability in the corona. Models of prominences have been produced which suggest that stable equilibrium is possible for the required lifetimes.

The observed darkness of sunspots is believed to be caused by the inhibition of convection by a magnetic field. Simple radially symmetric models of sunspots can be constructed which provide a magnetic field structure similar to that which is observed at the solar surface. The spot models also appear to be stable provided they contain sufficient flux. They decay much more rapidly than they could through ohmic diffusion. Either a turbulent diffusion process is involved or the spots split up into smaller sub-units which can decay more rapidly.

Solar oscillations can be observed in sunspots. The normal 5-minute oscillations are suppressed in sunspots relative to the normal photosphere and this is probably caused by the effect of the magnetic field on the wave propagation. There is an additional set of higher frequency oscillations in the chromosphere of spots and these oscillations have non-linear properties.

6 Activity in other stars

Introduction

If the Sun were the only active star known, it would be difficult to expect to get a full understanding of the causes of its activity. In fact, it is now possible to observe a wide variety of active stars. This means that their activity can be correlated with such things as their spectra and luminosity classes, ages, evolutionary state (pre-main-sequence, main sequence, post-main-sequence), rotation velocity, membership of a binary system and possession of an outer convection zone. By studying a wide variety of stars, it should be possible both to identify the causes of activity and to determine how activity varies during a star's life.

As has already been explained, most of the evidence for solar activity would not be observable even on a nearby star that happened to be just like the Sun. Because the disks of other stars cannot be resolved, surface features such as spots would not be apparent and neither would the small luminosity variations which occur during a solar cycle. Similarly the change in the bolometric luminosity of the Sun produced by even an intense flare would be insignificant. Some indications of activity might be picked up by the detection of chromospheric and coronal emissions in a wavelength range, where it is not totally swamped by the radiation from the photosphere, but this would not provide much of a basis for an understanding of stellar activity.

It is, however, a priori unlikely that we are living in a solar system associated with the most active star in the Galaxy. It therefore makes sense to look for stars which show activity similar to that of the Sun but in a more extreme form. Such stars have been discovered and it is therefore possible to use the correlations mentioned in the first paragraph of this chapter to try to understand the origin and evolution of stellar activity.

Before I discuss the results, I describe some of the types of activity which have been discovered in other stars and how they have been discovered. An early identification was that of flare stars. These are dwarf stars of spectral type M; the notation dMe is used where the letter e indicates strong spectral features in emission. They were originally found as stars with large erratic variations in their visual luminosity, which were attributed to flare-like phenomena much more intense than solar flares. Solar flares are characterized by non-thermal radio emission and X-rays and radio emission from the flare stars of spectral type M indicates that a phenomenon similar to but much more extreme than solar flares is probably being observed.

As has been mentioned earlier, magnetically active regions of the Sun can be observed in emission from the spectral lines of singly ionised calcium known as H and K. The study of H and K emission from other stars and of its variation with time was for a long time the principal indication of magnetic activity in these stars. For example, it allowed the tentative identification of phenomena similar to the solar cycle in stars for which there was no possibility of detecting sunspots like those on the Sun. Studies of pre-main sequence stars and of main sequence stars in young and intermediate age star clusters have indicated that magnetic activity is related to stellar age, as is stellar rotation velocity, whose evolution is probably controlled by magnetic activity. Studies give clues to the properties of the solar system at the time that the planets were forming, a point to which I shall return in Chapter 8. It appears that the Sun was probably much more magnetically active in the past than it is now.

The properties of the outer solar atmosphere could only be studied during eclipses until the invention of the coronagraph but even then the subject only developed fully when telescopes could be put into space to study the ultraviolet radiation and X-rays, which do not reach the Earth's surface. Space observations have led to the discovery of hot or warm coronae around a wide range of cool stars. They have also indicated that there is a division between those stars with hot coronae and a low rate of mass loss and those with cool coronae and very high rates of mass loss.

Stars have been observed to vary in luminosity in ways which cannot be understood in terms of intrinsic variability or eclipses in a binary system. The obvious explanation is that they have spots and that the variation occurs when the spots pass into or out of view as a result of stellar rotation. If that is the case, the spots must cover a much greater fractional area of the spotted stars than sunspots do on the Sun.

Our knowledge of the Sun is as great as it is because there has been a dedicated group of solar observers and telescopes which have looked at the Sun almost continuously for a long time. As a simple example observations over decades were needed to establish the existence and properties of the solar cycle and the Maunder minimum would not have been discovered without observations going back for centuries. Some properties of other active stars will also only be discovered as a result of regular observations made over a long time. This has the implication that much work must be concentrated on apparently bright stars because they can be observed using relatively small telescopes. It is unlikely that time will be regularly available on very large telescopes for a programme of monitoring the properties of active stars.

Stellar activity and stellar age

There is evidence that the degree of activity in main sequence stars is dependent on the age of the star in the sense that young stars are more active than older stars. Before I discuss this I must explain how the ages of stars can be estimated. This can only be done

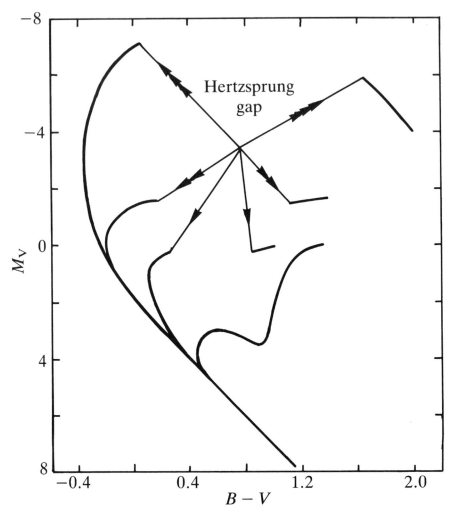

Fig. 38.   Schematic Hertzsprung–Russell diagrams of four galactic star clusters, showing in each case main sequence stars and red giants. The Hertzsprung gaps are regions where few stars are found. The youngest cluster has the highest turn-off from the main sequence.

directly in the case of stars which are members of star clusters. The method has been discussed in Chapter 6 of *The Stars* and will be repeated briefly here. For the present discussion I am only concerned with the relatively young galactic or open clusters in the disk of the Galaxy and not with the globular clusters in the halo. The Hertzsprung–Russell diagram of a set of galactic clusters appears schematically as in fig. 38. For each cluster there is a main sequence, a turn-off point from the main sequence and a group of red giants. The shape of the diagram is explained as follows. When main sequence stars have exhausted the hydrogen in their central regions, they turn into red giants. Because high mass stars have a very high luminosity relative to their mass, they evolve more

h and χ Persei	10^7
Pleiades	7×10^7
Praesepe, Hyades	4×10^8
NGC 752	10^9

Table 4. Approximate ages (in years) of young and intermediate age galactic clusters.

rapidly than main sequence stars. In any given cluster, the stars which are currently close to the turn-off point are the ones which are in the process of moving from the main sequence to the giant branch.

Calculations of stellar evolution, of the type which have been described in Chapter 3, enable the evolutionary tracks in the HR diagram to be obtained for stars of different masses and chemical compositions. Inside any star cluster it is assumed that all stars have essentially the same composition, so that only mass is important. The positions of stars in the HR diagram at any given cluster age define what is known as an *isochrone*. The age of the cluster is estimated by determining which isochrone gives the best fit to the observed cluster HR diagram. There are uncertainties of two types in this procedure. One involves the conversion of the observed stellar properties (apparent magnitude, spectral type) to the theoretical quantities (luminosity, effective temperature). The other involves the input data to the theoretical model, which have been discussed in Chapter 3, such as uncertainties in the theory of convection and in the detailed chemical composition of the stars.

The difficulties mentioned in the last paragraph affect estimates of the absolute ages of star clusters. It is therefore not surprising that different investigators obtain different values for these ages. What raises fewer problems is a determination of the relative ages of the clusters. There is for example no disagreement that the Hyades cluster is younger than the Pleiades even if there is doubt about their precise ages. Current estimates of the ages of some galactic clusters are given in Table 4.

The evidence of magnetic activity in cluster stars tends to come from the strength of their emission features in calcium H and K lines with the younger clusters having stronger emission and thus greater inferred magnetic activity. Having said that the aim is to see what other properties of the cluster stars change with age. This has a twofold purpose. The first is to help determine what is the underlying cause of stellar activity. The other is to identify properties which may enable the estimate of ages of stars which are not in clusters. Although the values of these ages cannot be expected to be as accurate as the cluster ages, they may be valuable in a statistical sense.

Stellar rotation and magnetic activity

The first property which does apparently vary with cluster age is the average rotation velocity of the stars. What can of course generally be measured is $v \sin i$, where i is the inclination of the axis of rotation to the line of sight. The average value of $v \sin i$ for young clusters is found to be greater than that for older clusters. This suggests that the rotation velocities of stars decline with age and it supports a link between stellar rotation and stellar activity. Why should a star rotate more slowly as it ages? Provided the star possesses both a magnetic field and a stellar wind, the process of magnetic braking, which has been discussed in Chapter 4, provides a natural explanation of the slowing down of rotation velocity. The inferred magnetic activity from the properties of the H and K lines is indirect evidence for the presence of the magnetic field. Obviously what I have first described is only a qualitative explanation of what may happen. It is necessary to determine whether plausible magnetic field strengths and mass loss rates do provide a braking time scale which is consistent with the observed change in the average value of v sin i. There have been some suggestions that initially only the outer convection zone is slowed down by the magnetic braking and that the core of the star responds on a larger time-scale. This might be possible if there is very slight coupling between the magnetic field in the convection zone and the deep interior of the star. In any discussion of the variation of stellar rotation rates from one cluster to another it cannot necessarily be assumed that their distribution of rotation rates at birth was broadly similar. Star formation remains an imperfectly understood subject and it remains unclear how the angular momentum of a star-forming cloud is communicated to the stars which form in it. This will be discussed further later in the chapter.

Lithium abundances

Another indication of stellar age is the lithium abundance in the atmosphere of a star with a deep outer convection zone. For relatively young clusters which are also reasonably close to the Sun, there may not have been any really major variations of lithium abundance in the gas out of which they formed and this should be comparable with the observed lithium abundance in the local interstellar medium today. The may in this sentence is important because lithium is so fragile. If it is nonetheless assumed that cluster stars formed with a similar fraction of lithium, the surface abundance should have changed with time. The reason for this is that lithium is readily destroyed by nuclear reactions in stellar interiors such as

$$^7\text{Li} + \text{p} \rightarrow {}^8\text{Be} \rightarrow 2\,{}^4\text{He} \tag{6.1}$$

and

$$^6\text{Li} + \text{p} \rightarrow {}^3\text{He} + {}^4\text{He} \tag{6.2}$$

and this destruction can occur at temperatures which are reached in the convection zones of late-type main sequence stars. This suggests that an estimate of the age of a star which is not in a cluster can be obtained from its lithium abundance.

There are several reasons for wishing to obtain a full understanding of stellar lithium abundances and at present the position is not completely clear. A very small amount of ^7Li is believed to have been produced in the early Universe. Because lithium is so easily destroyed in stellar interiors, it is believed that the oldest stars in the Galaxy, which are in the halo, should have a lithium abundance which is less than or equal to the amount produced cosmologically. How much less depends on the destruction mechanisms inside the stars. The deduced primordial abundance can provide a tight constraint on cosmological models.

Observations of lithium abundances in star clusters do indicate that there is a reduction in abundance with cluster age. In the case of a relatively young cluster like the Pleiades, there is no significant variation of abundance with the spectral type of the star being studied. In an older cluster, such as the Hyades, there is a clear depletion of lithium for stars with deep outer convection zones. In a really old galactic cluster M67, which is of similar age to the Sun, the depletion is greater and the Sun has an abundance similar to that of M67 stars of comparable surface temperature.

There has recently been a suggestion that the lithium depletion is not entirely caused by a deep outer convection zone but that there is an additional effect caused by what is known as rotational mixing. It is well known that it is not in general possible for a rotating star to be in precise hydrostatic equilibrium. There must be a slow circulation carrying matter around in the star and that circulation, in addition to the convective motions, could carry lithium to a depth where it could be destroyed. This could enhance lithium destruction in rapidly rotating young stars. The mixing processes in close binary systems are predicted to be different from those in single stars, because of tidal interactions which can cause the stars to be locked in synchronous rotation. It is predicted, and also observed in some cases, that such binaries could retain a higher lithium abundance than single stars of the same spectral type.

The possibility that lithium might be produced as well as being destroyed in active stars will be mentioned later in the chapter.

Influence of binary membership

It will be seen below that many of the most magnetically active stars are in binary systems and that some of the active stars are red giants. If they were single stars, their surface rotation speeds would have been expected to be significantly lower than those of young main sequence stars for two reasons. The first is that they would have been substantially reduced by magnetic braking because giants will generally be older than the main sequence stars; this is not totally clear because a giant will have been of earlier spectral type when it was main sequence star, so that it will have had a shorter main sequence lifetime and a shallower convection zone. The second reason is that the expansion to giant size with conservation of angular momentum reduces the surface angular velocity. These effects are counteracted in binary systems for the reason already mentioned in Chapter 2. Gravitational interaction between stars in a close binary can cause the orbital period and the rotation period to become the same and this means that these binary giants both are rapidly rotating and have a deep outer convection zone. This is a good recipe for magnetic activity.

After these introductory remarks, I discuss a variety of types of magnetic activity in turn. Before doing so I must stress that the study of stellar activity is a young and very rapidly developing field. This is particularly true because of the extension of observations into those wavelength ranges which are not accessible from the Earth's surface. It is certain that much of what I write will be out of date by the time that the book appears. I shall therefore concentrate on principles rather than on detailed results.

Starspots

Starspots cannot be observed directly but, if they are sufficiently large, they can be observed through their effect on the observed luminosity of a star. In the case of the Sun, the radiation emitted from the centre of a sunspot provides a luminosity per unit area which is between a fifth and a quarter of that from unit area of the normal photosphere; to obtain this value I have assumed that the emission is proportional to T^4 which is precisely true for black body radiation. This implies that, if starspots have similar properties to sunspots, the fractional area covered by spots must be somewhat greater than the fractional reduction in luminosity. It is inferred that starspots are present, if the luminosity of a star changes with time, and if this change does not have any obvious explanation such as intrinsic variability or eclipses in a binary system. As a star rotates a large spot may move in and out of sight and this leads to the observed luminosity change. It is irrelevant to this observation whether the presence of starspots leads to an increase in total stellar luminosity, as appears to be true for the Sun, or to a decrease, because in either case the observed luminosity of the star will be lower when the spot is facing the observer than when it is not. Of course, if there are periods when it is spotted and periods when it is not, any effect on the total stellar luminosity may become apparent. As will become clear such a study may require a very long timescale of observations.

Limb darkening and foreshortening

Before I discuss the observations of spots, it is necessary to introduce an idea known as *limb darkening*. This can only be observed readily in the case of the Sun and indirectly for eclipsing binary stars. The visible disk of the Sun does not appear uniformly bright, with the radiation from the edge or limb being less intense than that from the centre. The radiation reaching the observer from the solar limb has left the Sun at a large angle to the vertical. Radiation moving in the vertical direction is more intense than radiation moving at an angle to the vertical and as a result the edge of the disk appears darker than the centre. A not quite rigorous explanation of this phenomenon is illustrated in fig. 39. Radiation which reaches an observer from the Sun (or any star) travels from a layer in the Sun from which there is about a 50 per cent chance that it will escape without further absorption. Radiation from higher up escapes essentially freely and radiation from lower down is absorbed. The layer from which radiation escapes is the visible photosphere.

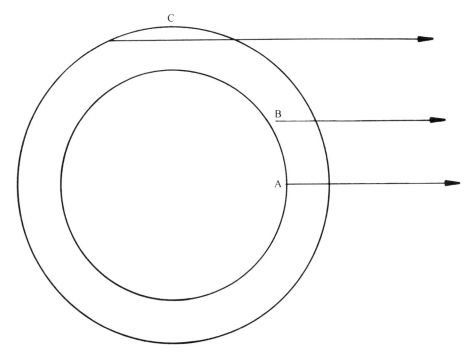

Fig. 39. Limb darkening. Light travelling to an observer from the centre of a stellar disk comes on average from the spherical layer passing through A. Light from B, which is at a lower temperature than A, can also just reach the observer; it travels through more absorbing material than if it were at the same level in the disk centre. At C radiation can pass right through the limb, but it all comes from matter at a lower temperature than at either A or B. As a consequence the stellar disk looks brighter at the centre than at the limb.

Because radiation travelling at an angle to the vertical passes through more solar material, for such radiation the photosphere is nearer to the solar surface. As the temperature is lower there, the observed radiation is less intense.

Limb darkening means that a totally black spot would subtract a different fraction of the observed luminosity depending on its position on the observed surface of the star but there is a further more important effect, foreshortening (fig. 40). A circular spot is assumed present on the stellar equator with the rotation axis perpendicular to the line of sight from the observer. The fraction of the observed surface that the spot covers will change as the star rotates. If the surface were uniformly bright, the fractional reduction in stellar luminosity would vary during the rotation cycle roughly as shown in fig. 41.

Determination of spot properties

It is not easy to use photometric observations alone to provide information about the position and location of spots. In the case of a star like the Sun seen at a distance the luminosity variations would be too small. Even if a star were much more spotted than

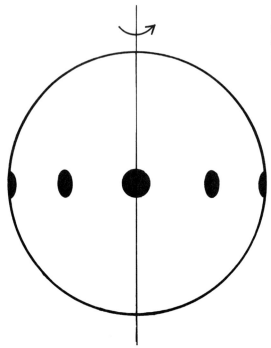

Fig. 40. Foreshortening. The fractional area of the stellar disk occupied by a circular spot varies schematically as shown as the star rotates, because of projection on to the line of sight.

the Sun but the spots were many and well scattered in latitude and longitude, there would be a strong cancellation effect and the overall luminosity variations in a single rotation period would be quite small. In this case the luminosity change due to spots might become apparent if the star possessed an activity cycle like the sunspot cycle so that there were periods in which there were very few spots. However, there would be no

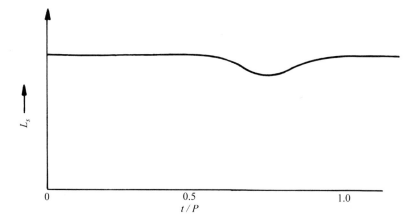

Fig. 41. A single spot affects the total observed luminosity of a star schematically as shown. For half the rotation period, P, the spot is invisible and for the second half the reduction in luminosity is related to the appearance of the spot in fig. 40.

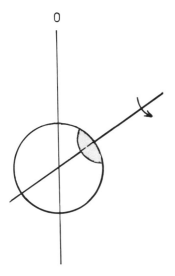

0

Fig. 42. Polar spots. If a star has a spot symmetrically placed about a pole, the appearance of the star as seen by an observer O does not change as the star rotates.

reason to observe such a star at all, let alone for an activity cycle of many years, unless it had other strong indications of activity. Clearly starspots will be most readily studied if a star has a small number of large spots because the full effect of the luminosity variations, or something close to it, will be apparent. Even in this case there will be problems for stars whose rotation axes are close to pointing towards the observer when spots might be visible throughout the rotation period and for polar spots in any case (fig. 42). Because of the possibility of cancellation of the effects of different spots, it is easy to underestimate the degree of spottedness of a star from photometric observations alone.

If observations are made in two different wavelength ranges, such as B and V of the UBV system, some further information is available. As explained in Chapter 2, the difference B–V, known as a colour index, is an approximate measure of the surface temperature of a star. If a star has spots of lower surface temperature than the normal photosphere, the colour of the spot is different from the rest of the surface and the value of B–V for the whole star will change as the spot moves in and out of sight. This means that an attempt can be made to model the properties of the spots using the observed colour variations of a star as well as the luminosity variations in one colour.

If a spot is observed on a single star of well-defined spectral type and luminosity class, the following calculation is possible. For such a star an estimate of its radius can be made either from equation (2.4), given that one has an idea of both its luminosity and effective temperature, or from theoretical models of such stars. An observation of the spectrum of the star provides a value of $v \sin i$, where v is the rotational velocity of the star and i is the inclination of the star's axis of rotation to the line of sight. From the radius and $v \sin i$, $P/\sin i$ can be calculated, when P is the rotation period. If the spot can be assumed to be fixed on the stellar photosphere or to be moving slowly compared to the rotational speed of the star, the observed light variations give a value for P. Comparison of $P/\sin i$ and P provides a value for i if the deduced value of $\sin i$ is less than unity; alternatively they suggest problems with the model. Clearly any attempt to determine the positions of spots on a stellar surface is greatly helped if a value of i is known.

Doppler imaging

There is another way in which the presence of spots can be deduced. Because the photosphere is cooler in a spot than outside, the properties of an absorption spectral line from a spotted region are different than those from a point on the normal photosphere. The distinct spectral line profiles from different parts of the photosphere cannot be observed for a star other than the Sun but the spectral line broadening provides limited spatial resolution across the stellar disk. Thus the region of the star which is moving towards the observer provides the shortest wavelength part of the spectral line, while the part moving away from the observer is responsible for the longest wavelengths. As the star rotates and a spot first becomes visible near the limb of the star, moves across the disk and then disappears at the other limb, the line profile changes as the radiation from the spot (provided it is large enough) is successively responsible for different parts of the line profile. The effect of the presence of a spot on a line profile is shown schematically in fig. 43. Note that this immediately gives some information about the latitude of the spot if it is assumed that the star rotates approximately uniformly. An equatorial spot can affect the whole broadened spectral line in turn. A spot nearer to the pole is never moving towards or away from the observer with the highest speeds represented in the line profile. A spot actually at the pole only affects the shape of the centre of the spectral line.

There are limitations to the information which can be obtained from such spectral lines. If a star were not rotating the spectral line would be broadened for other reasons. Sources of broadening include those due to thermal and convective motions in the photosphere as well as broadening produced by finite resolution of the instrument/ detector system. As a result a spot will affect more of the spectral line profile than its size would suggest. There is also the effect caused by the finite time of observation during which the spot's influence moves through the line profile. This effect is more important

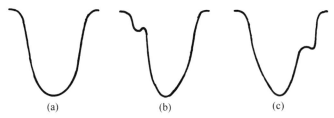

(a)   (b)   (c)

Fig. 43.   In the absence of a spot an absorption spectral line appears as in (a). If a spot is present it is cooler than the surrounding photosphere and it produces a shallower spectral line. Because the observed spectral line is from the entire stellar disk, the spot produces a perturbation in the line profile which moves through the line as the star rotates and the spot is moving successively away from the observer and towards the observer. This is shown schematically in (b) and (c).

for rapidly rotating stars and for fainter stars for which a longer exposure is required. It is also necessary for the influence of the starspot on the line profile to be large enough to be distinguished from other observational scatter. For all of these reasons the most reliable information is provided by large spots.

There are two approaches to an explanation of an observed line profile. The simplest approach in principle is to calculate the spectral lines produced by a wide variety of spots of different size, shape, darkness and position on a stellar disk and to compare these calculations with the observed line profiles. Although this method is straightforward in principle, there are so many free parameters that it is extremely time-consuming. As a result the usual approach is that of *image inversion* or reconstruction. This is a complex procedure and I shall not attempt to describe it in detail. I shall just illustrate the principle. The observed spectral line contains contributions from the entire stellar disk and there is a time series of observations showing how the spectral line changes as the star rotates. If a grid of points is distributed on the stellar surface and they are assigned arbitrary values of temperature, say, a synthetic spectral line can be calculated. The synthetic data is then compared with the observed data using a goodness of fit parameter and the synthetic data are then varied until a best fit with the observations is obtained. This method is found to be quite successful at determining the positions of spots, at least as far as longitude is concerned, but not so good at determining spot temperatures. These are best obtained by use of the photometric techniques discussed earlier. Because spotted stars are cool, observations in the visible and infrared (V, R, I) are most appropriate rather than the B–V mentioned earlier.

Properties of spotted stars

The best known spotted stars are the RS Canum Venaticorum (RS CVn) and BY Draconis (BY Dra) stars, each named after the first star of the class identified. The RS CVn stars and some of the BY Dra stars are binaries. As mentioned earlier the significance of the binary nature is that coupling between two components in a fairly close binary causes the two stars to rotate rapidly. As a result the cool component exhibits a high degree of magnetic activity. I shall discuss the relation between rotation and magnetic activity later in this chapter. These active stars are then no different in principle from very rapidly rotating single stars except that their rapid rotation is being maintained by the binary interaction, whereas single stars of a similar age might have slowed down as a result of magnetic braking.

RS CVn stars are all binaries in which the more active star is usually a K type subgiant or giant. The companion star is usually a main sequence star of earlier spectral type. Stars are put in this class if at least one star in the system has strong Ca II H and K emission and if there are periodic light variations which cannot be attributed to pulsation, eclipses or to distortion of the shape of a star by its companion. BY Dra stars are late-type K or M main sequence stars. They have strong hydrogen Balmer α in emission and they are active flare stars. Both types of stars have strong emission in the ultraviolet and X-ray spectral regions and this is an indication of stellar activity.

Although there are ambiguities in the interpretation of the observations, there is a general agreement that the fractional area of RS CVn stars that is covered in spots can vary between 10 per cent and 40 per cent or even higher compared with no more than 1 per cent on the Sun. The difference in temperature between the spotted areas and the normal photosphere is of order 1100 K which is significantly lower than that in the Sun; however it must be remembered that the normal photospheric temperature is also significantly lower. There is at least a suggestion that starspots are primarily found at higher latitudes in the Sun, individual spots are large and long-lived and it seems possible that stars have active longitude belts in which stars are usually found.

The observed properties of BY Dra stars are somewhat similar. Observations of BY Dra itself suggest spots both near the poles and in lower latitudes and up to 1992 spots had been continuously located in its polar regions for 14 years. Similar persistence of activity was found in another star EV Lac. Spot lifetimes may be several hundred rotation periods. Although these stars are more rapidly rotating than the Sun, these lifetimes are nevertheless much longer than sunspot lifetimes.

Another much studied single active star is the young main sequence star AB Doradus. I shall discuss some of its other properties later. Here I can say that observations at intervals of about a year for several years have shown that AB Dor possesses a spotted region near to the visible pole and also further spots at lower latitudes. The Doppler imaging technique, which I described very briefly above, indicates that the latter spots have been visible in approximately the same longitudes for years. As AB Dor has a rotation period of 12.4 hr, this is also evidence of persistence of magnetic activity for many rotation periods. Specifically recently published observations have shown that spots in low to intermediate latitudes appeared very similar in January 1992 and December 1992, although there was some change in structure of a polar spot. There was around 5 per cent of the area which was spotted. Although most of the spot system was stable for a long time, the star shows fluctuations in light output on a timescale of a week and this is also the lifetime of features in an extensive prominence system observed on AB Dor. This suggests that surface and coronal magnetic flux changes must be occurring on such a timescale and the properties of the polar spot may be related to this.

There has been some controversy about the existence of polar spots. Originally it was assumed that starspots would be likely to have the same low latitude distribution as sunspots. As observational evidence for polar spots accumulated, theoretical arguments were advanced in favour of the occurrence of polar spots in rapidly rotating stars. If rotation is strong, rising elements may find it very difficult to exchange sufficient angular momentum to move outwards in a radial direction and may be more likely to conserve angular momentum and move towards the pole. There is another reason for dispute about the existence of polar spots. As mentioned earlier they show their presence by a change in the shape of a spectral line, which does not change as the star rotates. As they are always visible, they also do not produce a rotational modulation in the light output from the star. A convincing detection of a polar spot requires a demonstration that no

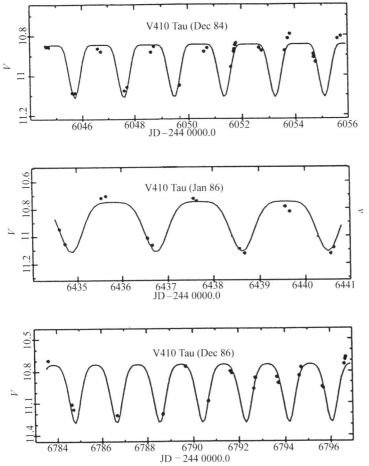

Fig. 44. *V* light curves for V410 Tau from Bouvier, J & Bertout, C, *Astronomy and Astrophysics*, **211**, 99, 1989. The filled circles are the observations. The curves are synthetic light curves obtained by assuming that the variations are due to starspots.

other source of spectral line broadening can produce the observed line shape. Observers are now convinced that this is the case.

Another group of stars which appears to have starspot activity is the pre-main-sequence T Tauri stars. They have a rotational modulation of luminosity which can be attributed to spots. The presence of the spots is confirmed by light-curve synthesis and they are comparable in their fractional size and overall behaviour with the spots on RS CVn stars. Figure 44 shows some observations of the star V410 Tau which can be understood if about half of the surface area of the star was covered by spots. I make some further remarks about T Tauri stars later in the chapter.

Interpretation and comparison with the Sun

It appears that a large fraction of the area of some spotted stars is occupied by spots. Because the stars show other evidence characteristic of magnetic activity it is assumed that the spots are produced where magnetic flux tubes break through the stellar surface. Because the differences in temperature between the spots and the surroundings are lower than in the Sun and because the photospheric densities of subgiants and giants are lower than that of the Sun, it appears that the characteristic magnetic field of starspots should be lower than that of sunspots. This conclusion is reached by a simple-minded application of the equation of pressure balance

$$B^2/2\mu_0 = nk(T_e - T_i),\tag{6.3}$$

where T_e and T_i are the temperatures inside and outside the spots and n is the particle number density. Thus there need not be greatly more magnetic flux breaking through the surface than there is on the Sun. A more detailed discussion of structure of the stellar surface would be needed to confirm this result.

To the best of my knowledge it has not yet been possible to obtain a direct measurement of the magnetic field of a spotted star. Even if it were possible, it would provide some average of the surface field rather than the field of a spot or spots. It is also not possible to be sure that an apparently large spot is in fact a single spot in which the field is all of one polarity rather than a closely packed concentration of smaller spots. Given the long lifetime of spots or spot groups, there is perhaps a suggestion that a single large spot of one polarity or a bipolar pair is more likely than a closely packed concentration of smaller spot pairs. It seems likely that such a configuration would evolve through successive reconnection to a much simpler configuration (fig. 45) There is however some evidence, that I will mention later, that the magnetic properties of the spotted regions do fluctuate while the overall spottedness remains. The subgiants and giants have higher luminosities as well as lower surface densities than main sequence stars of the same spectral type. If most of their luminosity is to be carried by convection the convective velocities will need to be more vigorous.

There is, as mentioned earlier, a strong suggestion that many observed starspots are concentrated at high latitudes whereas sunspots are preferentially at low latitudes. Some

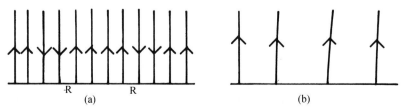

Fig. 45. A tight bundle of small spots, with an overall dominance of one polarity can evolve by reconnection in the regions marked R in (a) to the single large spot shown in (b) (in reality the configuration should be two-dimensional).

evidence has also been presented that starspots sometimes drift in latitude to the poles rather than to the equator, as in the case of sunspots. It is also quite likely that the stars' mean light is lower when spot activity is high in opposition to what happened in the Maunder minimum. It appears that dynamo models for the maintenance of stellar magnetic fields must not always give the same results.

The highly spotted stars have deeper convection zones than the Sun and they tend to have higher convective velocities. They are also mainly much more rapidly rotating than the Sun. The observation of long-lived large polar spots is at least suggestive that the rotational influence on the observed magnetic field structure is much stronger than in the Sun, as has already been mentioned above. It must however be stressed that the observed spotted stars are exceptional. Most subgiants and supergiants are slowly rotating and most lower main sequence stars may have a magnetic structure much more like that of the Sun, which is not observable in the way that I have been describing.

Emission line activity

Most stars, which are believed to show magnetic activity, are neither highly spotted nor have a directly measured magnetic field. It is fortunate that there is another way in which magnetic activity can be detected. In the solar chromosphere the Ca II H and K lines are observed in emission and it is found that the strength of the emission varies with the solar cycle being greatest at solar maximum. Since the early 1960s this emission has been studied in stars other than the Sun and it is now generally agreed that the strength of the H and K emission is a good indicator of magnetic activity. Once one has this property, one can then attempt to correlate it with other properties of the stars concerned, which are conjectured to be associated with or responsible for the magnetic activity. The correlations which have been found are very suggestive but their limitations must also be stressed. The strength of the dynamo action is believed to be determined by the dynamo number, D, for which an approximate expression is

$$D = \tau_c^2 / P_{rot}^2, \tag{6.4}$$

where τ_c is a characteristic convective turnover time and P_{rot} is the rotation period of the star.

The dynamo number for any particular star is estimated from a spherical model of that star using a model for the properties of the convection zone – specifically the mixing length theory of convection with a value of the mixing length which appears appropriate – and it is assumed that the important dynamo action is located near to the bottom of the convection zone or possibly in the convective overshoot region, which connects the outer convection zone to the radiative interior. When this estimate is made for main sequence stars and early post-main-sequence stars, a reasonably good correlation between magnetic activity and dynamo number is found (fig. 46). There is some scatter, which may be due to inadequate theoretical ideas, but which should not be unexpected when we consider what we know about the Sun. We must recognise that the dynamo number as

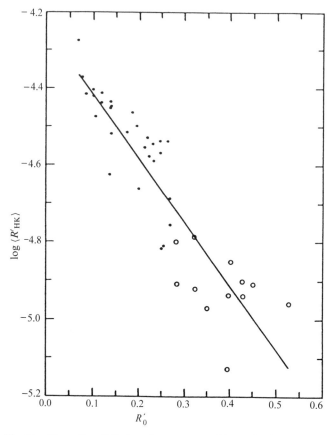

Fig. 46. Magnetic activity in main sequence and immediately post-main-sequence stars from Gilliland, R L, *Astrophysical Journal*, **299**, 286, 1985 based on observations by Noyes, R W *et al.*, *Astrophysical Journal*, **279**, 763, 1984. R'_{HK} is a measure of Ca II H and K activity and R'_0, is $(P_{rot}/\tau_c)(l/R_c)^{1/2}$ where P_{rot}, τ_c, l and R_c are respectively the rotation period, convective turnover time, mixing length and radius at the base of the convection zone. Filled circles are young stars and open circles are old stars.

defined above, which depends on average properties of the stellar exterior, would have been the same during the Maunder minimum, when H and K activity was presumably lower, as it is now. Thus any plot of solar values during the last few hundred years in such a diagram would show a large scatter.

There has also been a study of stellar activity in other late-type stars, with a search for evidence for periodic activity similar to the solar cycle. Clearly the monitoring of stars for decades is required before unambiguous evidence can be obtained but useful information has now been accumulated. An interesting tentative result is that stars with weak emission like the Sun appear to have magnetic cycles with periods of the order of a decade, whereas the more active stars tend to show smaller variations in their chromospheric activity. There are exceptions to this result, but if it is generally true it suggests a

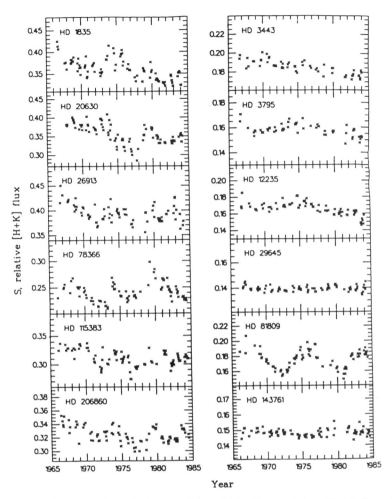

Fig. 47. Stellar activity cycles in stars of about 1 $M_\odot$. (By permission of S L Baliunas.) The left panel shows young stars and the right panel old stars.

difference in the pattern of convection with level of activity. Observations of chromospheric activity in stars of about a solar mass are shown in fig. 47. The stars in the right hand column are most like the Sun in mass and age. It is interesting that most of them do not have variations of as high an amplitude as the Sun. The stars in the left hand column are younger and exhibit more irregular variations.

There is some corresponding evidence for a correlation between activity and dynamo number in pre-main-sequence stars but here the scatter is much greater as is shown in fig. 48. I have suggested a factor which should be considered in attempting to interpret this result. It is not known to what extent pre-main-sequence stars may retain an interior fossil magnetic field which is a relic of their formation phase. If they do, the strength of the field may differ significantly from star to star depending on the particular

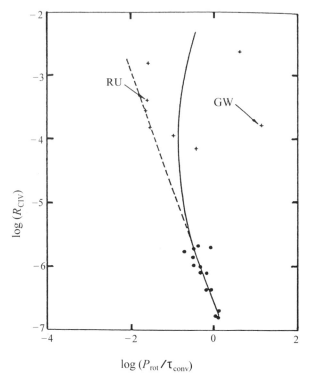

Fig. 48. Magnetic activity (as measured by emission in C IV) in pre-main-sequence stars from Tayler, R J, *Monthly Notices Royal Astronomical Society*, **227**, 533, 1987 based on Gilliland, R L, *Astrophysical Journal*, **300**, 339, 1986. Crosses are pre-main-sequence stars and filled circles main sequence stars. The solid line is a fit to the observations and the dashed line is an extrapolation of the main sequence relation. Two particularly studied stars are indicated.

circumstances of the star's formation including its angular momentum. If the star as it approaches the main sequence has a convection zone, which occupies essentially all of the star as is predicted by the simplest theoretical models, the length scale of the fossil field can be reduced by the convective motions and eventually the fossil field will be subject to ohmic decay. At any time the field can be considered partly fossil and partly dynamo and magnetic activity may be higher in those stars where the fossil field dominates. By the time the main sequence is reached, the fossil field has disappeared and the activity is determined entirely by the dynamo field.

Evolution of stellar rotation

As has already been mentioned, one property of the Sun and other late-type stars, which is closely related to the magnetic activity, is the evolution of the angular velocity of the star. If a logical progression of angular velocity with time for stars younger than the Sun can be established, that will provide useful information in considering the past history not only of the Sun but also of the solar system. Provided that any mass loss is occurring from the star, the interaction between the outflowing matter and the magnetic field leads to magnetic braking of the rotation. The outflowing matter corotates with the magnetic field structure, while the magnetic field controls the flow, and the angular momentum which it carries away is more than its initial share of the star's angular momentum. As the degree of magnetic activity, including the overall field strength, is believed to be determined by the stellar rotation, the braking should be strongest when the rotation is most rapid. In an early discussion it was shown that both the rotation frequency and the Ca II emission decayed approximately according to a $t^{-1/2}$ law, where t is the time since the star reached the main sequence; because the most rapid braking occurs at early times, the precise time at which this behaviour starts is not crucially important. As a result, one might expect to have only a relatively small spread in properties of stars of the same mass and age, once they were well settled on the main sequence, regardless of their initial rotation.

There has always been a problem in understanding stellar rotation. The density of a star is more than 20 orders of magnitude greater than the interstellar gas out of which stars form. It is impossible for stars to form with conservation of angular momentum but even without complete angular momentum conservation it might be expected that newly formed stars would be very rapidly rotating. It was therefore a surprise when it was found that pre-main-sequence T Tauri stars are relatively slowly rotating, although there is a significant spread in their rotation velocities. It appears that the loss of angular momentum from protostar must occur very early in the pre-main-sequence history of a star, so that the angular momentum problem associated with star formation is solved before the T Tauri phase.

When the rotation velocities of T Tauri stars are compared with those of the young main sequence stars which they are going to become, it appears that there must be a rapid spin-up with conservation of angular momentum as the star contracts to the main sequence, after which the magnetic braking of the type which has been discussed above occurs. One critical question is how rapidly the braking of the outer layers is communicated to the interior of the star. If the transfer of angular momentum from the convective exterior to the radiative interior is slower than the spin-down rate, this could be very important. In that case young main sequence stars could have cores rotating more rapidly than their surfaces. Although the helioseismological observations do not allow very substantial non-uniform rotation in the Sun at the present time, this does not tell us what the Sun was like in the past.

The original discussions of magnetic braking adopted a very simple model of the magnetic field structure of a rotating star. It subsequently became clear from studies of the solar corona that a distinction must be made between those magnetic field lines which pass through coronal holes and well out into interplanetary space and those which form closed loops in the corona. The mass loss and thus the magnetic braking is restricted to the open field lines. The tendency of rapid rotation to lead to a stronger magnetic field, which can enforce corotation out to a larger radius and hence lead to increased braking is to some extent counterbalanced by an increased fraction of closed loops. The overall effect of improvements in the simple model is to lead to a braking rate which is comparable with but somewhat stronger than the $t^{-1/2}$ law quoted earlier.

Stellar flares

As I have explained already, solar flares would not be apparent if the Sun was at the distance of even a relatively nearby star. In disk integrated optical light the Sun is certainly not a flare star. The ACRIM experiment on the Solar Maximum Mission (SMM) satellite did not see even the most intense solar flares, which produced at most a 0.01 per cent change in the solar irradiance. It was however discovered in 1948 that there are stars which have very intense flares. The first such star observed had the spectral classification dMe, where d stands for dwarf (or main sequence) and the e means that the star's spectrum contains emission lines. This star had a 1 magnitude outburst in the optical, which means that its optical luminosity increased by a factor of about two and a half. Intense flares are observed on a large number of K and M dwarfs. Because these stars have much lower effective temperatures than the Sun, their optical light is a much smaller fraction of their total luminosity, which means that the change in bolometric luminosity is much less than the observed change in the visible luminosity. The flares are nonetheless very impressive and the largest ever observed is of 7 magnitudes or a factor of over 600. Typical intense white light flares have a luminosity of 2×10^{22} W and a total energy emission of more than 3×10^{24} J. At a given rate of occurrence optical flares on dMe stars are 10 to 10^3 times as energetic as solar flares. The intensity of these flares is even more remarkable when it is realised how much smaller M dwarfs are than the Sun. The smallest on which flares have been observed are comparable in size with Jupiter and many of them must have a surface area which is less than 10 per cent of the surface area of the Sun. The very intense flares must occupy a larger fraction of the star's surface than is the case for solar flares.

Although the low mass main sequence stars were the first to be discovered to have flares, the most intense stellar flares occur on the RS CVn binary stars, which have already been discussed in connection with starspots. It is not uncommon for such stars to have flares with an energy release of a few times 10^{29} J and the most energetic observed was in excess of 10^{31} J. All of the stellar flares are, like the solar flares, believed to be associated with the reconnection of magnetic field lines. In the case of the RS CVn binary, flare loops have been observed which almost extend from one star to the other. There are suggestions that the strongest flares may involve field lines which connect the two stars.

Radio flares

Flare stars were also the first genuine radio stars to be discovered, other than the Sun. The radio emission from many of the dMe stars is much greater than that from the Sun but once again the highest emisssions are associated with RS CVn binaries. Stellar radio flares are observed to be up to 10^5 times as luminous as solar radio flares. As was mentioned when the Sun was discussed, the radio emission is produced by energetic particles, which are accelerated by the electric fields generated in the flares. There does, however, appear to be a difference in the precise emission mechanism for the solar and stellar flares. Radio astronomers characterise the emission at any frequency in terms of something which they call the *brightness temperature*. This is the temperature which would be required in order for black body radiation to have the observed emission at that frequency. It is not unusual for cosmic radio emission to have a high brightness temperature, which means that it is produced by a non-thermal process. Brightness temperatures as high as 10^{16} K have been observed on some flare stars. It is believed that this emission must be *coherent*. In a coherent emission process, a large number of particles act together to produce a much more intense emission than they could produce individually. An example of what is meant by this is the following. An accelerated charged particle emits radiation proportional to q^2, where q is its charge. If N particles emit incoherently, the total emission is proportional to Nq^2. However if the particles can behave as if they were a single particle of charge Nq, the coherent emission will be proportional to N^2q^2, a very much greater emission, if N is large.

Nuclear reactions in flares

The radio emission from flares is believed to be produced by the energetic electrons. Although there are also believed to be a comparable number of energetic ions, they are not anything like so efficient at radiating energy; for a given particle energy, the emission tends to be inversely proportional to m^2, where m is the particle mass. The energetic ions can however collide and this may lead to nuclear transmutation. At the highest flare temperatures, there is some possibility that *thermonuclear reactions* might lead to some conversion of deuterium to helium. There is not much deuterium present, so that this cannot be a very important reaction. More significant could be the occurrence of *spallation nuclear reactions*. These are reactions which cause the breakdown of nuclei into smaller nuclei. Thus the most abundant nuclei of H and ^{4}He can collide with and break up other abundant nuclei such as ^{12}C, ^{14}N and ^{16}O. Such spallation reactions involving cosmic rays in the interstellar medium are believed to be the only significant method of producing the lithium, beryllium and boron found in the Universe, other than the ^{7}Li which was produced in the big bang. It is possible that some lithium is produced by spallation reactions in the surface regions of flare stars. If that is the case, it will need

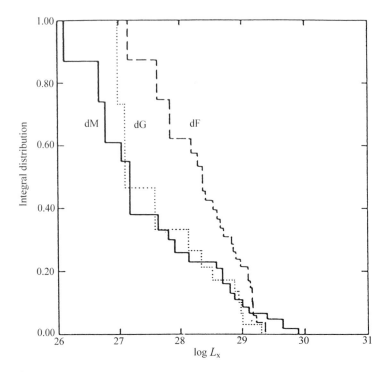

Fig. 49. X-ray emission from dwarf stars from Rosner *et al.* (1985). The X-ray luminosity is in erg s^{-1}. The Sun's emission places it at the bottom end of the measured values. Reproduced, with permission, from the *Annual Review of Astronomy and Astrophysics* Vol 23 © 1985 by Annual Reviews Inc.

to be taken into account in any discussion of the evolution of lithium abundance in low mass stars.

X-ray emission from active stars

The discussion so far refers to observations from the Earth's surface and most of the detailed observations of magnetically active stars which can be made in this manner are concerned with stars which are much more active than the Sun. The use of ground-based diagnostics such as Ca II H and K emission automatically biases the observations towards these more active stars. It is therefore very difficult to get a clear view of the entire range of solar type activity. To avoid this problem, it is necessary to transfer attention from the photosphere and chromosphere to the transition region and the corona. This means that observations should be made in the ultraviolet or X-ray regions of the spectrum, which in turn means from space. In this way it becomes possible to study a volume-limited sample of stars, because even quite a low level of coronal X-ray emission, for example, can be detected given that there is no corresponding photospheric emission.

When observations of a volume-limited sample of stars were made with the *Einstein* satellite, it was found that all dwarf stars with spectral types from dF to dM are X-ray emitters with luminosities in the range 10^{19} W to 10^{24} W. This is shown in fig. 49. This X-ray emission is evidence of coronal activity in all of these stars and the Sun is near to the bottom end in X-ray emission. Pre-main-sequence stars have an emission which can be 1000 times more vigorous than that of main sequence stars of the same spectral type, but this X-ray emission is not well-correlated with Hα activity. There is in contrast quite a good correlation between Hα activity and the X-ray luminosity, L_x, for main sequence stars including both dMe stars and the Sun. The later type main sequence stars, like the Sun, show irregular activity associated with flares as well as the steady emission of the quiet corona. An X-ray flare was observed on the star vB8, which is one of the faintest known M dwarfs. This has a bolometric luminosity of only about 5 × $10^{-4}L_\odot$ an inferred mass of about $0.1\,M_\odot$ and the spectral type dM7e. This observation raises a puzzle for dynamo theory, as will be discussed later.

T Tauri stars

I have mentioned T Tauri stars several times earlier in the chapter and now I will make a few comments summarising their properties. They are stars in the final stages of formation and approaching the main -sequence. It used to be believed that the processes of star formation and pre-main-sequence evolution were distinct but it is now clear that this is not the case, as will be discussed further in Chapter 8. An entire star does not form at once and accretion from a disk may be important even after a star first reaches the main sequence. It is certainly important in the T Tauri phase. T Tauri stars are normally divided into two groups, the *classical* T Tauri stars (CTTS) which were first discovered and which have strong Hα emission and *weak-line* T Tauri stars which are much weaker in Hα. I shall mainly describe CTTS.

These stars have strong Hα and Ca II H and K lines in emission. They have hot coronae and strong ultraviolet fluxes. As mentioned earlier there are quasi-periodicities in their light curves, with periods of a few days, which can be interpreted as rotational modulation due to spots. Typically the spots cover about 10 per cent of the surface area and they are present on WTTS as well as CTTS. The spots are typically 700 K to 1000 K cooler than the normal photosphere. Spot lifetimes can be very long. In the case of the WTTS V410Tau, a spot lifetime of a few years represents more than a thousand rotation periods. An equivalent lifetime on the Sun would be more than 80 years or almost four complete solar cycles. There are clearly important differences as well as similarities between the Sun and other active stars. The properties of T Tauri spots are very similar to those on RSCVn and BYDra stars. Some T Tauri stars in contrast show hot spots. This cannot be similar to solar activity and they are thought to be due to local heating when matter is accreted from a disk. For the normal cool spots, the inferred magnetic field strength is of order 0.1 to 0.3 T. This seems a reasonable figure for a young late-type active star.

T Tauri stars also have properties similar to the main sequence flare stars. They show optical outbursts which are 10^5 to 10^6 times as intense as a strong solar flare. They also have X-ray flares which are 10^5 times as strong as a strong solar flare. The X-ray flux and the $H\alpha$ flux are both anticorrelated with rotation period or equivalently correlated with rotation velocity, as has already been mentioned in my previous discussion of activity in pre-main-sequence stars. Because of the presence of disks in CTTS, there are some problems in separating what I am regarding as normal stellar activity from phenomena due to disk star interactions. In addition to disks, many T Tauri stars also show bipolar outflows of matter, as I shall mention further in Chapter 8, and this further complicates the interpretation of the observations.

In the past ten years there have been a number of attempts to measure the surface magnetic field of T Tauri stars. A first reported measurement of a field of about 0.06 T on the star was subsequently withdrawn as being unreliable, but a later measurement of a field of order 0.1 T on another star TAP35 now appears to be accepted.

Stellar coronae

When it became possible to study the high atmospheres of cool stars with the IUE satellite, there was a rapid increase of knowledge of coronae in stars other than the Sun. The results obtained with IUE were confirmed and extended by studies in the X-ray region with *Einstein* and *ROSAT*. An important discovery was that there was in general a division between the properties of the very late type giant stars and earlier type giant stars, which also have convective outer layers. Specifically a transition occurred at about spectral type K1. The stars to the left of this transition in the HR diagram (*i.e.* the hotter stars) have atmospheres like the Sun with very hot coronae. In contrast the cooler stars have high rates of mass loss in cool stellar winds and no hot coronae. This division in the giants does not extend to the more luminous supergiants, which show very strong mass loss.

In time it became clear that the division is not completely sharp and that there exist so-called *hybrid stars*, which have both hot material in their atmospheres and also high velocity winds. This is perhaps not surprising. As has been discussed in Chapters 2 and 5, the Sun possesses both a hot corona and a mass loss in the solar wind. A rate of mass loss similar to that of the Sun would not be observable in other stars, but it might be expected that there would be other stars which show this property in a more extreme form.

There is as yet no agreed explanation for the variety of properties in the atmospheres of cool stars. It is generally agreed that both the coronal heating and the mass loss must be driven by the magnetic fields. As the efficiency of a stellar dynamo is related to the rotation rate and as conservation of angular momentum implies a lower rotation rate in giants and supergiants than in main sequence stars, the outer layers of giants and supergiants should perhaps be different from those of main sequence stars. If mass loss leads to significant magnetic braking, the rotation rates can be reduced further. As stars become giants and then supergiants, the escape velocities from their surfaces become smaller and this makes it easier for mass loss to occur.

I have explained earlier that the hot solar corona is associated with magnetic field lines which are closed loops, which do not extend very far above the visible surface, whereas the solar wind flows along open magnetic field lines which extend far into the interplanetary medium. There is therefore a temptation to say that the stars with hot coronae have a preponderance of closed loops and that those with massive winds have a preponderance of open field lines. Even if that is true, it is not an answer to the question of the origin of the differences between the two classes of stars. Obviously the wind cannot escape if it is confined by closed magnetic field loops but the question is what came first. Did the strong wind drive out the magnetic field lines or did the field structure encourage the wind?

Evidence from star clusters

As I have mentioned earlier it is only for star clusters that there is a reasonably reliable estimate of stellar age. When a volume-limited group of stars is observed in some spectral range, such as the X-ray observations of M dwarfs, there is a spread of activity amongst stars of given spectral type. It is possible that a major cause of this activity spread is that the less active stars are older than the more active stars but it is not possible to check this because ages cannot be assigned to individual stars. It is therefore desirable to study the nearby star clusters in as much detail as is possible. Here it has to be stressed that cluster membership is not unambiguous. There are always stars in the same direction as the cluster which are nearer to us or further away. Very careful observations of their velocities and proper motions, of their parallaxes, if they are near enough, and of their magnitudes and colours are necessary to distinguish the most probable cluster members from foreground or background stars. If this can be done for two or more clusters, a study can be made of the dependence of stellar activity on age.

Such observations have been made for three clusters of increasing age, α Persei, the Pleiades and the Hyades. There is evidence that the colour at which Hα emission is first detectable is redder in the Hyades than in the Pleiades. In addition the colour at which every observed cluster member has Hα emission is redder in the Hyades. Equivalently this means that there is a range of stellar surface temperature for which Hα activity is absent in the older cluster, while it is present in the younger cluster. In addition stars in α Persei are on average more active than stars in the Pleiades of similar surface temperature. A recent study of the Hyades shows that the luminosity in Hα in the active stars is typically about 10^{-4} of the bolometric luminosity of the star, with a maximum value of $\sim 3 \times 10^{-4}$. For low luminosity stars, the ratio of the X-ray luminosity to the bolometric luminosity is more like 5×10^{-4} with a maximum of 10^{-3}.

In the Pleiades, the lowest luminosity stars from which X-rays have currently been detected have, in contrast, the significantly higher value of 10^{-2} for the same ratio, with a value of 10^{-3} being quite common for a wide range of stellar absolute luminosity. If one per cent of a star's total luminosity can be channelled into coronal X-rays, this requires an almost maximal level of magnetic activity. A significant fraction of the energy which is input near the photosphere must presumably be lost in other forms. There is, however, no tendency for the Pleiades stars to have more intense Hα emission. Indeed for the coolest ones the reverse appears to be true. The conclusion of a study of the Pleiades is that there is a large spread in activity in very low mass stars which may arise because of a spread in rotation velocities. At present there are no observations to test this idea.

In the Hyades, the spread of activity observed in stars of the same spectral type appears to be caused by a variation in the fraction of the stars covered by active areas rather than by a variation in strength of active areas of the same size. Observations of the stars at several epochs will be required to determine whether the same stars are always the more active or whether the observed scatter is a reflection of stellar activity cycles in which all of the stars share.

Activity in very low mass stars

Throughout this book I have expressed the view that activity in the Sun and other stars is governed by the magnetic field in the outer convection zones of these stars and that this magnetic field is maintained by a dynamo type process. I have also repeated a commonly held view that the main site of the dynamo is in the overshoot region just below the convection zone proper. In the case of the Sun, helioseismological observations show that there is a very strong velocity shear in this layer and this is believed to be good for the dynamo mechanism.

Models of stellar structure of the type that I have described in Chapter 3 suggest that below a certain mass main sequence stars should be fully convective in the sense that the region in which convection plays a role in energy transport extends throughout the entire star. In pre-main-sequence stars, convection is even stronger and more massive stars, as well as the very low mass stars, can be expected to be fully convective. If a star is fully convective, it cannot possess an overshoot region below its convection zone. It was therefore expected in advance of observations that there would be a reduction of activity in the stars of the latest spectral type. That is not what is observed. As mentioned earlier, the star vB8, which has almost the lowest possible mass of a main sequence star and which certainly must be fully convective if theoretical ideas are correct, has shown an intense X-ray flare. It appears that dynamo action must be possible in stars without an overshoot region, even if the overshoot region is very important in the Sun and similar stars.

A detailed study of the X-ray emission from all nearby low mass stars has shown a trend of reduced X-ray luminosities for the fully convective stars compared with the slightly more massive stars which have radiative cores. It is not, however, obvious whether this is anything more than what would arise because the surface areas of the fully convective stars are smaller than those of their more massive counterparts. This study was based on observations taken with the *Einstein* satellite. A later satellite, *ROSAT*, has higher sensitivity and it has provided evidence which is not in full agreement with the Einstein results and which suggests that the magnetic heating of stellar coronae is not sensitive to changes in the internal structure of very low mass main sequence stars. Study has been made of vB10 a star comparable in mass with vB8. This is in a binary system with a somewhat more massive companion and this provides an opportunity to compare the properties of two stars of different mass and internal structure but of the same age. It appears that vB10 uses at least as much of its bolometric luminosity in coronal heating as does its companion, and this reinforces the need to discover a dynamo model which is effective in the absence of a convective overshoot region. Further study of very low mass stars will be necessary before the situation becomes completely clear.

The observations of the Pleiades, which I mentioned earlier, provide rather inconclusive evidence about activity in very low mass stars. Some of the highest X-ray emission in relation to bolometric luminosity comes from very low mass stars which must be assumed to be fully convective. On the other hand, there is a wide spread in activity levels in such low mass stars. This suggests that while some fully convective stars can have dynamos which are at least as efficient as those in stars with an overshoot layer and a convective core, on average fully convective stars have less efficient dynamos.

Activity in high mass stars?

So far all of my discussion of stellar activity has referred to stars with outer convection zones, with the assumption that the activity is produced by a dynamo-generated magnetic field whose properties change with time. Recently X-ray emission has been detected from some early-type stars of spectral type Ae and Be; as before the letter e indicates that these stars have emission lines in their spectra. The X-ray emission suggests that these early-type stars, which certainly do not possess outer convection zones, have hot coronae. If they do, does this have any relation to the existence of coronae in late-type stars?

There is no reason why early-type stars should not have a magnetic field. In fact, the strongest observed magnetic fields in main sequence stars are in stars of spectral type A, the peculiar A stars. These magnetic fields are believed to be fossil fields rather than dynamo-generated fields. Such fields would not decay during the main sequence lifetime of the stars and there is no need for them to be regenerated. They are, however, believed to be fixed in the stars, with the only observed variations being produced because the star's magnetic axis does not coincide with its rotation axis. Such a static magnetic field should not stimulate any coronal heating or other manifestations of coronal activity.

A possible solution arises because the Ae and Be stars are rapidly rotating. It is possible, but by no means certain, that velocity gradients in the outer layers might drive turbulent motions and produce a dynamo field, whether or not a fossil field was already present. At present this idea must be regarded as rather speculative.

Prominence activity on AB Doradus

I have already mentioned AB Doradus as a cool young star with strong and persistent starspot activity. It is the nearest solar type star which has reached the main sequence very much later than the Sun. Various properties such as its rotation rate and lithium abundance suggest that it is of a similar age to the stars in the Pleiades. AB Dor was noted as a very strong emitter of X-rays. The X-ray spectrum implies the existence of a hot gas with a temperature of about 2×10^7 K filling a volume which is very much larger than that of the star. Closed coronal magnetic field loops extending to two or three stellar radii above the photosphere must be filled with the hot gas. At such large radii, the centrifugal force due to the rotation of the star has a very important influence on the equilibrium of the loops. Thus, in the stellar equatorial plane, there is a balance between centrifugal force and gravity at the Keplerian corotation radius which satisfies the equation

$$GM/r^2 = \Omega^2 r, \tag{6.5}$$

where M and Ω are the stellar mass and angular velocity. For AB Dor, the corotation radius is about 1.36 stellar radii.

The X-ray observations were followed by the discovery of an interesting structure to the Hα spectral line of AB Dor. The line shows transient absorption features, which show a rotational modulation. These can be interpreted as being produced by dense clouds of cool hydrogen (the other chemical elements are of course also present) high in the corona of the star, which are forced to rotate with the star because they are associated with the magnetic loops. In the original observations they appeared to form preferentially between three and four stellar radii from the star's surface and then to move radially outwards. Because they form outside the corotation radius, the influence of centrifugal force must exceed that of gravity unless they are very close to the star's pole. Later observations indicated that the clouds actually form much nearer to but still outside the corotation radius at a rate of one or two a day. Further out the clouds are observed to be moving in excess of 30 kms^{-1} away from the star. These clouds are gigantic *prominences* which are very much larger than anything that is observed on the Sun. The outward moving clouds could lead to a substantial mass loss and also to magnetic braking in a period of order 10^8 yr.

The mass loss in these prominences is very different from the solar wind mass loss though perhaps not from the coronal mass ejections. In the Sun the closed magnetic loops containing the hot corona are situated well within the corotation radius. It appears that in younger stars the more rapid rotation brings the corotation radius closer to the stellar surface while at the same time it leads to a higher level of stellar activity with a higher input of energy to the corona. By the time that AB Dor is as old as the Sun, its properties should be much more similar to those of the present Sun, although not the same because it will continue to be of later spectral type.

Some theoretical studies have been made of coronal loop structure in an attempt to understand why the cool prominences form. Equilibrium models of coronal loops on rapidly rotating stars show a change of structure if the loop summit lies outside the Keplerian corotation radius, where the centrifugal force is stronger than gravity. This outward net force, which must be balanced by the magnetic force, leads to a density increase in the upper parts of the loop. It is argued that this increase in density leads to enhanced cooling of the loop plasma and to a thermal instability producing prominence-like condensations of neutral material just outside the corotation radius.

Although AB Dor has been the most studied star because it is the nearest and brightest of its type, it is obvious that there should be many more young stars showing such prominences. Prominence activity has been found in other nearby G-dwarfs in the young α Persei cluster.

I have been able in this chapter to give nothing more than a very brief summary of a large and rapidly increasing amount of information about the properties of active stars. Although the featureless spherical models of the stars discussed in *The Stars* provide a very good description of the global and average properties of stars, it is now clear that a very wide variety of stars have very complex outer layers. It is no longer true that all stars look uninteresting at a distance of ten parsecs. Much of the surface activity is not of great relevance to the evolution of the star but, if it drives substantial mass loss, that is very important for stellar evolution.

Summary of Chapter 6

Because the Sun is very close to us, its activity is very apparent, but this would not be true even if the Sun were one of the other nearest stars to us. A study of stellar activity has therefore relied on the existence of stars, whose activity is very much greater than that of the Sun. In all cases it is believed that the activity is driven by a dynamo-generated magnetic field in the outer layers of the star, but in general no direct measurement of the magnetic field strength has yet proved possible.

Observations indicate that stellar activity is related to stellar age in the sense that the younger stars are the more active ones. Because the Sun is an old star, it does not have to be thought of as one of intrinsically low activity. The more active but younger stars of a similar spectral type to the Sun will probably become like it as they age. Stellar age can be determined for members of star clusters. The properties of cluster stars then provide indicators for age in individual stars.

Of course age is not the true defining parameter for activity. The underlying property which changes with age is stellar rotation and it is the rotation of the stellar convection zone, which is supposed to power the dynamo which generates the magnetic field. Rotation is believed to decline as a result of magnetic braking associated with mass loss. The very high activity of some binary stars can be understood because they are kept in synchronous and rapid rotation even as they age.

Observed stellar activity includes starspots, whose presence can be deduced from a rotational modulation of stellar luminosity and spectral line profiles. Stellar spots tend to be long-lived and they can cover a very much larger fraction of a stellar surface than do sunspots. Some starspots are found near the poles suggesting that dynamos do not always behave in the same manner. An early indication of stellar activity was emission in the H and K lines of singly ionised calcium. Observations show a correlation of activity with rotation. There is some evidence for stellar cycles similar to the solar cycle, although in most cases observations have not been made for long enough.

Stellar flares can be very much more intense than solar flares both in absolute energy release and even more so as a fraction of the stellar luminosity. The first genuine radio stars discovered, other than the Sun, were the dMe flare stars. X-ray emission is also very intense during flares but it also occurs when the stars are quiescent and it is one of the pieces of evidence for hot coronae. In the case of giant stars there is a division, but with exceptions, between the very late-type giants, which have intense stellar winds and cool coronae, and earlier type giants, which have coronae more like that of the Sun.

Models of the dynamo production of magnetic fields tend to have the main site of the dynamo in the overshoot region just below the convection zone. Very low mass main sequence stars are believed to be fully convective and do not have an overshoot region. It is therefore something of a surprise that stellar activity persists to the lowest mass main sequence stars which have been observed and this is one of many problems which needs to be resolved.

One star, AB Doradus has been observed to have intense prominence-type activity high in its corona. Cool clouds and gas are observed to be condensing and moving outwards from the star leading to significant mass loss. This can only be studied in detail because AB Dor is the nearest young solar-type star to us.

7 Solar–terrestrial relations

Introduction

In this chapter and the next one I move outwards from the Sun to consider its relation with the solar system. In this chapter I concentrate specifically on the present interaction between the Sun and the Earth, with a few comments on other planets, and in Chapter 8 I discuss the origin of the solar system.

It is trivially true that without the Sun there would be no Earth as we know it. The luminosity of the Sun would not have to change very much in either direction for the Earth to be unable to support its present flora and fauna and density of human population. There is currently some controversy about the influence of the Sun on terrestrial weather, in particular whether there is a correlation between weather and sunspot number. There is, however, no doubt that weather on the Earth is almost entirely determined by the interaction of solar radiation with the outer layers of the Earth. That the weather should be so complicated and unpredictable, even when the radiation of the Sun is essentially unchanging, is a consequence of such things as the ellipticity of the Earth's orbit, the Earth's rotation, the inclination of its axis of rotation to its orbital plane and to the division of the Earth's surface between sea and land and to the existence of mountain ranges.

It is not this interaction of the steady Sun with the Earth that is my primary concern in this chapter. Instead I am concerned with the manner in which erratic changes on the solar surface, which have been discussed in Chapter 5, have consequences on the Earth and with the way in which direct exploration of the Earth's environment by spacecraft is providing an understanding of the interaction. The principal intermediary in this is not the electromagnetic radiation of the Sun but is the corpuscular emission through the solar wind, which has already been discussed in Chapters 2 and 5. The Earth's environment is not only of interest for its own sake but it proves to be a valuable laboratory for the study of properties of ionised gases or plasmas, in a way which is not possible on Earth.

Before the theoretical prediction of the existence of the solar wind and its subsequent discovery by spacecraft experiments, it was believed that interplanetary space was essentially empty most of the time. It was recognised that the observed small bodies in the solar system, asteroids, comets, meterorites, meteors, must be accompanied by even smaller particles of interplanetary dust, for which there was direct observational evidence close to the Earth, particularly in the properties of the zodiacal light, but there was no indication of a significant amount of interplanetary gas. It had been suggested that there was an ejection of ionised gas from the Sun at the time of a solar flare and that observed changes in the Earth's magnetic field known as *magnetic storms* were produced by these particles. It also appeared that some cosmic rays were produced at the same time.

The existence of the solar wind implied a new view of the steady-state relationship between the Sun and the Earth. Previously it was only necessary to consider the interaction of photons with the atmosphere and surface of the Earth. Through this interaction the outer layers of the atmosphere are ionised by high energy photons from the Sun and the resulting *ionosphere* is primarily the concern of those interested in radio communication. The lower atmosphere is not ionised and it is here that the weather systems originate, so that it is the concern of meteorologists. The solar wind introduces a new region known as the *magnetosphere*, which lies beyond the ionosphere.

The magnetosphere

The solar wind particles moving outwards from the Sun carry lines of solar magnetic field with them as has been explained in Chapter 5. As they approach the Earth, they must become aware of the existence of the Earth's magnetic field. As I have discussed in Chapter 4, charged particles cannot readily move across a magnetic field and as a result either the magnetic field controls the motion of the charged particles or the particles move the magnetic field around. When the solar wind is far from the Earth, the kinetic energy density of the solar wind greatly exceeds the energy density of the Earth's magnetic field and the solar wind therefore pushes the magnetic field towards the surface of the Earth. Eventually, this is no longer true and the Earth's field is able to control the motion of the solar wind.

If I make the approximation that there is a uniform spherically symmetric solar wind, the Earth is a small obstacle to its motion. The Earth's magnetic field increases the size of the obstacle and it forces the solar wind to move round the Earth at some significant distance from the solid surface, with the Earth's magnetic field being trapped within the magnetospheric cavity which is shown schematically in fig. 50. To a first approximation, on the solar side of the Earth the cavity boundary is situated where the kinetic energy density of the solar wind is equal to the energy density of the Earth's magnetic field or

$$\frac{1}{2}\rho v^2 = \frac{B^2}{2\mu_0}, \tag{7.1}$$

where ρ and v are the solar wind density and velocity and B is the Earth's magnetic induction. In applying equation (7.1) it should really be recognised that the value of B has been increased above what it would be if the solar wind did not exist, because all the magnetic field lines outside the magnetospheric boundary have been confined within it

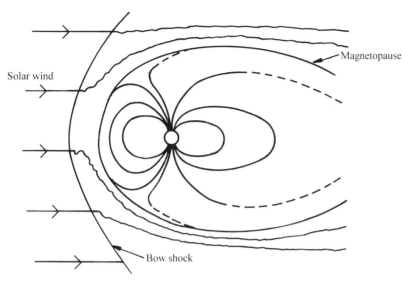

Fig. 50. A schematic picture of the magnetospheric cavity, showing in particular the incoming solar wind, the bow shock, the magnetopause and the trapped terrestrial magnetic field. At the bow shock, solar wind plasma is slowed down and some of its energy is turned into random motions indicated by the wiggly lines beyond the shock.

by the solar wind. As a dipole magnetic field decreases like $1/r^3$, this is not however a great effect.

As can be seen in fig. 50, the magnetosphere is very far from being spherical. There is a long *magnetotail* on the far side of the Earth from the Sun. In addition there is a lack of symmetry because the Earth's rotation axis (and its magnetic axis, which currently departs slightly from the rotation axis) is not perpendicular to the line from the Sun. In addition, even if the solar wind moves radially outwards from the Sun, the solar magnetic field lines which it carries with it are inclined to that direction as I shall explain later in the chapter. Inside the magnetospheric cavity, the Earth rotates every 24 hours, and the whole system rotates about the Sun annually. At the same time the Sun rotates about once a month. Even if there were no erratic changes in the solar wind's properties, the solar rotation would lead to changes in the magnetosphere simply because the solar wind does not travel outwards in a spherically symmetric manner.

In this chapter I am concerned with two things; how spacecraft are providing an understanding of the properties of the Earth's immediate environment and what happens when there is a sudden change in the properties of the solar surface and of the solar wind. Before I can do this I must introduce some ideas from *plasma physics*.

Plasma frequency and plasma oscillations

The name plasma was given to an ionised gas in the 1930s when it was realised that a gas could sustain coherent oscillations at a frequency known as the *plasma frequency*. The system was thus supposed to wobble like a jelly or plasma. If a plasma of pure hydrogen is considered, the expression for the plasma frequency ω_p is

$$\omega_p^2 = \frac{Ne^2}{\epsilon_0 m_e} + \frac{Ne^2}{\epsilon_0 m_p}, \tag{7.2}$$

where N is the number of electrons (or protons) per unit volume and the other symbols have their usual meanings. Because $m_p \gg m_e$, there is very little difference between the value of ω_p and that of the electron plasma frequency ω_{pe} defined by

$$\omega_{pe}^2 = Ne^2/\epsilon_0 m_e. \tag{7.3}$$

The same is also true for an electrically neutral gas composed of a mixture of electrons and ions of various charges. For that reason I shall not distinguish between ω_p and ω_{pe} in what follows. The expression for ω_{pe} is derived in Appendix 8.

The existence of the plasma frequency has an important effect on the propagation of electromagnetic waves. If the electromagnetic wave has properties which vary like $\exp(ikx-i\omega t)$, the wave number k and frequency ω usually satisfy the relation

$$\omega = kc, \tag{7.4}$$

if propagation is in a vacuum. If instead it propagates in an ionised gas, this relation is replaced by

$$\omega^2 = \omega_p^2 + k^2 c^2 \simeq \omega_{pe}^2 + k^2 c^2. \tag{7.5}$$

From equation (7.5) it can be seen that both the phase velocity (ω/k) and the group velocity ($d\omega/dk$) of the wave differ from c; the group velocity is the speed at which information propagates and this is always less than c.

It can also be seen that, if $\omega < \omega_p$, k is imaginary. This means that waves cannot propagate. This leads to the reflection of low frequency waves by the Earth's ionosphere as is shown in fig. 51. This makes possible long wave radio communication around the

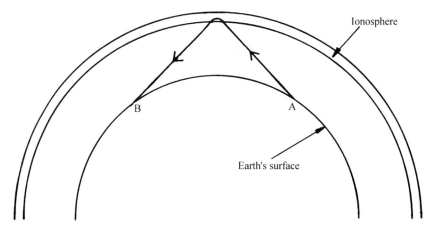

Fig. 51.   The reflection of radio waves from the ionosphere. A wave of frequency ω emitted from A is reflected from the level in the ionosphere where the plasma frequency ω_p is equal to ω.

Earth. It also produces a natural cut-off frequency for radio astronomy from the Earth's surface, because waves from cosmic sources with $\omega < \omega_p$ are reflected from the top of the ionosphere. If there is any significant energy at such low radio frequencies, it will only be able to be studied using radio telescopes above the ionosphere. Note that, as the ionosphere is maintained by solar radiation, its properties above any point on the Earth's surface vary from day to night. Equation (7.5) is also derived in Appendix 8.

Because almost all the known matter in the Universe is in the form of a plasma, which can be regarded as a fourth state of matter in addition to solids, liquids and ordinary gases, it might be thought that coherent plasma phenomena would be extremely important. Thus the entire Sun is composed of a plasma even if the photospheric layer is only lightly ionised. Coherent plasma phenomena are not always important because in my discussion of plasma oscillations I have assumed that the electrons are acted on by no forces other than the electric field. I have also assumed that thermal motions of the electrons are unimportant and that is not always the case as is discussed in Appendix 8. Another obvious effect that has been ignored is that there are collisions between electrons and of electrons with ions. If the frequency of collisions is higher than ω_{pe}, collisions will disturb the ordered motion of the electrons which is necessary to maintain the coherent oscillation. I now give a very crude estimate of the collision frequency.

The collision frequency

I can estimate the electron/electron collision frequency as follows. Two electrons will influence one another significantly, if (see fig. 52) at their distance of closest approach, b, their mutual electrostatic potential energy exceeds their kinetic energy. Thus approximately

$$e^2/4\pi\epsilon_0 b > m_e v^2/2. \tag{7.6}$$

All particles passing closer than the value of b, for which the inequality is replaced by equality,

$$b_{crit} = e^2/2\pi\epsilon_0 m_e v^2, \tag{7.7}$$

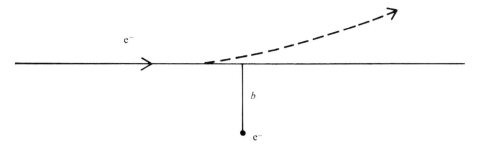

Fig. 52. Electron/electron collision. An incoming electron whose distance of closest approach to a stationary electron would be b is deflected by electrostatic repulsion on to the dashed path.

will be said to *collide*. A particle will behave as if it has a cross-sectional area $\sigma = \pi b_{\text{crit}}^2$ and, if it travels with speed v through a distribution of n_e electrons m^{-3}, the mean time between collisions τ_c is given by

$$\tau_c = 1/n_e \sigma v, \tag{7.8}$$

or equivalently the collision frequency is

$$\nu_c = n_e \sigma v = n_e e^4 / 4\pi \epsilon_o^2 m_e^2 v^3. \tag{7.9}$$

If the electrons form a thermal distribution at temperature T_e (which may differ from the ion temperature T_i), a typical electron velocity will satisfy

$$v^2 = 3kT_e/m_e. \tag{7.10}$$

If I insert this, I obtain for a typical electron collision frequency

$$\nu_c \approx n_e e^4 / 4\pi \epsilon_0^4 m_e^{1/2} (3kT_e)^{3/2}. \tag{7.11}$$

This is only a very approximate discussion and the numerical coefficient is not exact. It is however good enough to obtain an approximate value of ν_c. If $\omega_{\text{pe}} \gg \nu_c$, the plasma oscillations will not be affected by collisions. If $\nu_c \gg \omega_{\text{pe}}$, collisions will destroy coherent oscillations. This is true in the solar interior and in the interiors of other stars. This dependence of collision frequency on density and temperature has been used in chapter 4, where I explained that the conductivity of an ionised gas is proportional to $T^{3/2}$.

The Debye length

In my discussion of plasma oscillations I assumed that a charge distribution of equal numbers of protons and electrons produced no net charge and no electric field. This cannot of course be strictly true. The particles are not a smooth distribution of charge but are effectively point particles. Around every particle there is a region in which there are no other particles and in which there is an electric field due to that one charge; I have already assumed that in discussing the collision frequency. As I move further away from the charge, I shall meet other particles and I can ask how far I need to move before I have lost all knowledge of the presence of the charge.

A charge $+e$ in a vacuum has the Coulomb potential

$$\Phi = e/4\pi\epsilon_0 r. \tag{7.12}$$

In a plasma electrons moving around at typical thermal speeds v_{th} ($v_{\text{th}} \approx (3kT_e/m_e)^{1/2}$) tend to be attracted by $+e$, while protons are repelled, so that on average in the vicinity of $+e$ there is a slight excess of negative charge. This is known as *Debye screening*. It reduces Φ and the screened potential can be shown to be

$$\Phi = (e/4\pi\epsilon_0 r)\exp(-r/\lambda_D), \tag{7.13}$$

where λ_D is called the *Debye length*. The value of λ_D is found to be

$$\lambda_D = (\epsilon_0 k T_e / n_e e^2)^{1/2} \tag{7.14}$$

It is thus approximately equal to v_{th}/ω_{pe}.

In order for coherent plasma oscillations to be possible Debye screening must be effective. There must therefore be a large number of electrons within a volume λ_D^3. Thus I require $n_e \lambda_D^3 \gg 1$. This ensures that a particle interacts via Coulomb interactions with many other particles simultaneously. A further comment should be made on the discussion of plasma oscillations which I have given. I assumed that all of the electrons were at rest initially. In fact they will be moving with their thermal velocities. As a result of that, plasma oscillations will be disturbed if the particles can move through a wavelength λ ($= 2\pi/k$) of the oscillations in a time of order ω_{pe}^{-1}. This indicates that for coherent plasma oscillations I require

$$\lambda \geq \lambda_D. \tag{7.15}$$

The wavelength must exceed the Debye length.

In diffuse plasmas, such as those in the interplanetary medium, coherent plasma oscillations can occur. Collisions are very infrequent and we have what is known as a *collisionless plasma*.

Measurements in space

Although much information about the Earth's interaction with solar radiation and the solar wind has been and is being obtained from ground-based observations, our understanding has been transformed by the presence of spacecraft capable of making *in situ* measurements of the properties of the magnetospheric and interplanetary plasmas. Spacecraft carry a large number of different instruments. Measurements can be made of all the components of electric and magnetic fields, of the bulk velocity of the ionised gas and, for example, of the distribution of particle velocities relative to the bulk velocity or equivalently the distribution of particle energies. If the particles have a well-defined temperature, this distribution should be the Maxwellian distribution of velocities. However there is frequently not a well-defined temperature because it is collisions which cause a particle velocity distribution to approach the Maxwellian form and, as I have explained, collisions are frequently unimportant. Even if there is some approximation to a temperature, it is often found that ions and electrons have different temperatures.

If an experiment is done in a laboratory plasma, some sort of probe is often pushed into the plasma so as to make a measurement. In designing the experiment, it is necessary to ensure as far as possible that the probe does not seriously disturb what one is trying to measure. If, for example, the probe carried an electric charge, it would not be a suitable device for measuring the electric field which was present in its absence. A satellite is a large object which is bound to disturb its immediate environment. It is necessary to ensure that any measurements that are made are not related to the presence of the

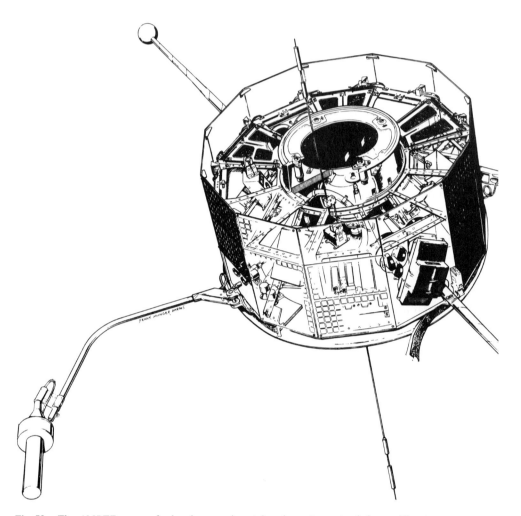

Fig. 53. The *AMPTE* spacecraft, showing experimental equipment on extended arms (Courtesy Munger, F and the Rutherford Appleton Laboratory).

satellite or to its properties. There is always a particular problem in interpreting measurements of electric fields because of the Galilean transformation law of electric fields. If we move from one frame to another moving relative to it with velocity **v**, there is the relation

$$\mathbf{E}' = \mathbf{E} + \mathbf{v} \times \mathbf{B}. \tag{7.16}$$

If there is a magnetic field, it is impossible to say that there is no electric field present. An induced electric field is always found if the frame of reference is changed.

It is, in fact, easy to see that, if a spacecraft is sitting in a plasma in which the electrons and ions have the same temperature, the surface of the spacecraft will acquire a negative charge. There are equal numbers of positive and negative charges but, because of equipartition of energy, the negative charges will have higher speeds and the flux $n_e v_e$ of electrons hitting the spacecraft will exceed the flux of ions. There will then be what is called a positive ion sheath with a thickness of about a Debye length outside of which the spacecraft will have a zero net charge. If any piece of equipment is to be placed in a part of the medium that is not affected by the sheath that surrounds the space vehicle, it must be held, on a support that projects from the body of the vehicle, at a distance greater than the sheath distance. The need to place instruments away from any disturbance produced by the spacecraft leads to them being put on long arms as shown in fig 53.

Spacecraft orbits

Ideally, if we are studying the properties of interplanetary space between the Sun and the Earth, we would like to obtain a full three-dimensional study of the system. Clearly this is impossible with one spacecraft or even with a finite number of spacecraft in simultaneous operation. There are two approaches to a space experiment. The most common one involves placing a satellite in a stable orbit. If the Earth were a point mass, the orbit of the satellite would be a closed ellipse, so that a given satellite would sample only a very limited region of the Earth's environment. There would be a choice between a circular orbit, which would have the satellite's velocity and distance from the Earth related by

$$\frac{v^2}{r} = \frac{GM_E}{r^2},$$ (7.17)

where M_E is the mass of the Earth, and a more elongated orbit, which would sample regions further from the Earth. Note that the size of the orbit determines the orbital period and for orbital radii not much greater than the Earth's radius this is about $1\frac{1}{2}$ hr. Because the Earth is not exactly spherical, the elliptical orbits do not close and this means that an individual satellite samples a larger volume of space.

There is one particular problem in interpreting measurements made in space. A spacecraft may record a variation with time in the plasma properties. What is not immediately clear is whether this variation arises because the spacecraft has moved into a region with different plasma properties or whether the properties would have changed even if it had been stationary in space. In many cases what is observed is a combination of these two effects. It is for this reason that the complex CLUSTER mission was planned for late 1995. The launch date has now been moved in 1996. If the launch is successful, it will be fully operational before this book is published*. It will employ four spacecraft whose relative positions can be varied. It is hoped that their simultaneous observations will help to unravel the spatial and temporal properties of the magneto-spheric and interplanetary plasma.

---

* CLUSTER was destroyed on launch on 4 June 1996 so that the planned mission will not ocur.

The purpose of CLUSTER is to investigate three-dimensional space plasma phenomena which are believed to occur throughout space but which are most readily studied in the neighbourhood of the Earth. Such near-Earth plasmas have small scale structures with spatial extent of several hundreds to several thousands of kilometres or equivalently a few to a few tens of particle gyration radii. As mentioned above, single satellite measurements suffer from an intrinsic inability to distinguish unambiguously between spatial and temporal variations. With two satellites this ambiguity is removed only for simple motions of essentially one dimensional structures. An unambiguous determination of three-dimensional structures requires a minimum of four spacecraft flying in a tetrahedral configuration and equipped with instruments capable of measuring fields and flows in three dimensions. CLUSTER will consist of four spacecraft in non-coplanar orbits. It will be possible to vary the separation of the spacecraft, which will study regions in the magnetosphere and in the solar wind. A typical orbit will be elliptical with dimensions $3\,r_E$ by $20\,r_E$.

The other experimental technique is to use a space probe. This is a spacecraft which travels through the solar system with no intention of going into a stable orbit about the Earth or any other body. Good examples of probes are the Mariner shots to Mars, Giotto to Comet Halley and the Voyager missions to the outer planets. If such a probe carries instruments which are capable of studying fields and particles, some transitory information about space plasmas, particularly at large distances from the Earth can be obtained. Voyager was able to study the magnetosphere of Jupiter and its electromagnetic interactions with its satellite Io directly. Sometimes an experiment shares both aspects. Thus after Giotto had visited Comet Halley, it was placed in a parking orbit from which it was moved to study Comet Grigg– Skjellerup in 1992.

Most space physics measurements are *passive* in the sense that they just measure what is there. In that sense they cannot be regarded as true experiments. There are some measurements which are *active*. An example of this is the release of a cloud of gas to produce an artificial aurora. If charged particles follow magnetic field lines, one way to study the topology of magnetic fields is to release charged particles and to observe to where they migrate. If this is to work, they must be able to be identified and observed. This is similar to the use of dyes in fluid dynamics experiments where the transport and dispersal of a dye may be the simplest way to study the fluid motions.

If a satellite is placed well above the Earth's atmosphere, it will have a very long life in a stable orbit. Its usefulness will depend on the lifetime of its instruments, or the ability to keep its orientation well-defined so that the geometry of the measurements can be understood and on the maintenance of a radio link between the spacecraft and the ground stations so that the experimental results can be retrieved. The same considerations apply to astronomical satellites. The ultraviolet satellite, IUE, whose results have been discussed in earlier chapters, has been in orbit since 1978 and is still operational at the time of writing, because it has proved possible to maintain its pointing accuracy much longer than was expected.

With the geophysical missions, the transmission of data to Earth is a major problem.

With all the instruments on the spacecraft, more information can be collected than can be transmitted to Earth down a radio link in real time. As a result data compression techniques are required to get the information to Earth. This will be a particularly extreme problem with the four CLUSTER spacecraft. Once it has been received the analysis of data frequently continues for years after a mission ends. The data analysis side of a mission is complementary to and as important as the hardware design. For most satellites more than one ground station is required, if the information is to be transmitted as it is collected, because there is no part of the Earth's surface which is always visible from the satellite. This is not the case if the satellite is placed just at the radius where the satellite's rotation period equals the Earth's rotation period and the satellite is in or close to the equatorial plane of the Earth. Note that it is easiest to launch a satellite from close to the Earth's equator and in that plane because the velocity possessed by virtue of the Earth's rotation helps in achieving the orbital velocity. Most space missions have been near to the Earth's equator or to the ecliptic, to which the Earth's axis is inclined at $23\frac{1}{2}°$.

After this general introduction to the properties of plasmas and of space missions, I now return to a discussion of solar–terrestrial relations.

Magnetic storms

Historically the first serious studies of solar-terrestrial relations came from a study of magnetic storms at the Earth's surface. These tend to be associated with and follow periods of intense activity on the surface of the Sun. There is a sudden change in the value of the magnetic field at the Earth's surface (sudden commencement) followed by fluctuations and a gradual return to the pre-storm state (fig. 54). In the 1930s S Chapman and V C A Ferraro noted that, in order to account for the delay between the time of activity on the Sun and the occurrence of a magnetic storm, the interaction between the Sun and the Earth must involve a stream of particles rather than solar radiation, which would travel at the speed of light. They therefore suggested that there was corpuscular emission from the solar surface at the times of great activity. As I have already explained, the prediction and detection of the solar wind altered this picture. There is a continuous flow of matter from the Sun which produces the magnetospheric cavity around the Earth. It is alterations in the structure of the magnetosphere, when the Sun is very active and when the solar wind becomes more intense, which produces the magnetic storm. I will discuss magnetic storms in slightly more detail when I have discussed some of the other magnetospheric phenomena.

van Allen radiation belts

The solar wind was one of the first discoveries of space missions. Another early discovery was that of the van Allen radiation belts. Spacecraft moving out from the Earth passed

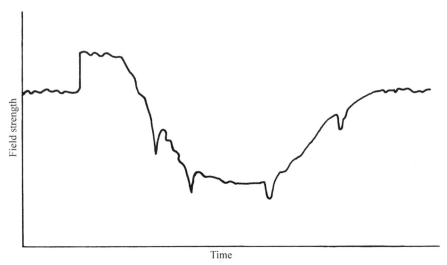

Fig. 54. Schematic picture of the variation of the magnetic field at the Earth's surface during a magnetic storm. The sudden commencement and enhanced field is followed by a period of increased ring current and reduced field with local oscillations superimposed. Eventually the field recovers to its pre-storm value.

through a region where there was an enhanced density of energetic charged particles. It appeared that the particles must be trapped for a long time in the region where they are situated. Three questions then arise. Where do they come from? How are they trapped? How do they eventually escape? The radiation belts were first discovered by the Explorer I spacecraft in 1958. The early experiments were designed to detect high-energy particles, > 30 MeV protons and > 1.6 MeV electrons. Observations of these particles showed two radiation belts, one between the ionosphere and $2 r_E$, where r_E is the Earth's radius, and the other from about $3 r_E$ to more than $4 r_E$. This double peak distribution led to the idea of two radiation zones. Subsequent measurements of lower energy particles showed that the radiation stretches all the way from the ionosphere to the magnetopause (fig. 55).

Magnetic mirrors and velocity space loss cones

The trapping mechanism is believed to be the *magnetic mirror* mechanism. This was an early idea for the confinement of a hot plasma in a magnetic field, which was suggested as a mechanism for producing controlled thermonuclear reactions in a laboratory. If a charged particle moves in a magnetic field which does not vary with time, its kinetic energy is constant. Thus

$$v_\perp^2 + v_\parallel^2 = \text{constant}, \tag{7.18}$$

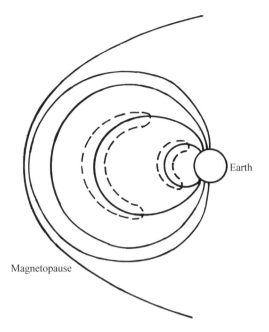

Fig. 55. Radiation belts are now known to lie along all the magnetic field lines between the Earth and the magnetopause, with the more energetic protons and electrons both being nearer to the Earth. The dashed curves show regions where particles with a particular band of energies might be found.

Earth

Magnetopause

where $v_\perp$ and $v_\parallel$ are the components of velocity perpendicular to and parallel to the magnetic field. In addition a quantity known as the *magnetic moment* of charged particles is almost precisely constant. Thus

$$\mu \equiv mv_\perp^2/2B \approx \text{constant.} \tag{7.19}$$

μ is what is known as an adiabatic invariant. Provided that the spatial variations of B are sufficiently slow, the departure of μ from constancy only becomes apparent after an extremely large number of orbital periods of the particle around the magnetic field. The constancy of the kinetic energy and of μ is discussed in Appendix 9.

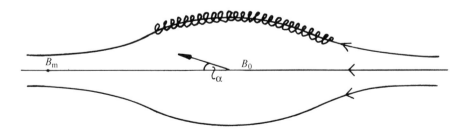

B_m
B_0
α

Fig. 56. The helical motion of a charged particle in a mirroring magnetic field. If the pitch angle, α, of the particle's motion in the centre of the mirror, where $B = B_0$, is large enough, the particle's motion is reversed before the maximum field value, B_m, is reached. If $\alpha < \alpha_{0m}$ given by equation (7.22) the particle passes right through the mirror.

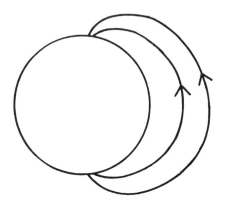

Fig. 57. A sketch of the Earth's magnetic field showing converging field lines towards the poles which enable particles to be trapped in a magnetic mirror.

Consider now the motion of a charged particle into a region of increasing magnetic field. As the value of B increases, so does $v_\perp^2$ according to equation (7.19). If the increase of B is sufficiently large, it is possible that a point will be reached at which $v_\parallel^2 = 0$ according to equation (7.18). At that point the particle must be reflected and move back into the region of lower B. A schematic magnetic mirror is shown in fig. 56. In an approximately dipolar magnetic field such as that of the Earth, magnetic mirroring is also possible because of the increase of the field towards the poles (fig. 57).

The problem with the magnetic mirror geometry is that particles with large $v_\parallel$ are not trapped. Return again to fig. 56. Let the magnetic induction at the centre of the mirror be B_0 and the maximum induction be B_m. Let the velocity components of a particle at the centre be $v_{\perp 0}$, $v_{\parallel 0}$. A particle will just be reflected if its perpendicular velocity in B_m is given by

$$v_{\perp m}^2 = v_{\perp 0}^2 + v_{\parallel 0}^2 \equiv v_0^2. \tag{7.20}$$

But

$$v_{\perp m}^2 = (B_m/B_0)v_{\perp 0}^2, \tag{7.21}$$

so that reflection will just occur if

$$\frac{v_{\perp 0}^2}{v_0^2} = \frac{B_0}{B_m} = \sin^2 \alpha_{0m}, \tag{7.22}$$

where α is the angle between the direction of motion and the axis of the mirror at the centre. α is called the *pitch angle* and for $\alpha < \alpha_{0m}$ particles pass right through the mirror without reflection.

This would raise no problem if the particles in the *loss-cone* (those with pitch angles less than α_{0m}; see fig. 56) simply escaped and left the others to be trapped for ever. If collisions are important, this will obviously not work because particles will continuously be scattered into the loss-cone. It is possible to choose parameters so that the mean time between collisions is longer than the desired trapping time but even this is not enough.

There prove to be other instabilities which scatter particles into the loss-cone and empty the mirror trap much more quickly than is required. It is a long time since magnetic mirror machines were considered a good approach to controlled thermonuclear fusion. A similar situation occurs in the magnetospheric plasma. At $r = 6\,r_E$, $B_0 \simeq 100\,\text{nT}$ while $B_m \simeq 0.5 \times 10^{-5}\,\text{T}$. The loss cone has α_{0m} of a few degrees only. Even so particles are scattered into it and *precipitate* into the ionosphere, often during auroral displays, and in the so-called *diffuse aurora*.

I have now discussed how energetic particles can in principle be trapped for very much longer than their time of travel along the magnetic field lines of the Earth and I have also discussed how they can escape as a result of instability. I now need to discuss how they are trapped in the first instance. It is clear that any particle which can enter a trap can also escape from it. It is no good relying on collisions to retain a particle once it is inside because they would equally effectively release trapped particles. There are only two possibilities, both of which were tried in laboratory experiments. The first involves opening up the trap to let particles in and then closing it behind them. The other involves injecting particles in a different state from their trapped state. I now discuss mechanisms of the latter type.

Suppose first that an energetic cosmic ray particle penetrates deeply into the Earth's atmosphere. It can do this because its gyration radius, which for a relativistically moving particle has the form E/eBc, where this E is the particle's energy and not the electric field strength, is very much larger than the length scale of the magnetospheric transition. As a result it is not confined by individual field lines. Suppose this particle collides with an atomic nucleus and ejects an energetic neutron. Some of the neutrons produced then travel into the magnetosphere and decay through the reaction

$$n \rightarrow p + e^- + \nu_e. \tag{7.23}$$

The disintegration energy plus the kinetic energy of the neutron can produce protons with energies of order 10 MeV and these protons can then be trapped.

Another suggestion is that energetic protons may sometimes be produced, during solar disturbances, by a stream of neutral hydrogen emitted from the Sun with great speed. It is supposed that, as it passes through the magnetosphere, a rapidly moving hydrogen atom sometimes picks up a charge from a (low energy) magnetospheric proton according to the reaction

$$H(\text{rapid}) + p \rightarrow p(\text{rapid}) + H, \tag{7.24}$$

and is transformed into an energetic proton. In addition, protons with energies of a few hundred MeV are sometimes emitted from solar disturbances (solar proton events). If they penetrate the magnetosphere, they can produce neutrons, and then energetic protons, in the same way as the galactic cosmic rays. The resulting protons have lower energies than those produced by galactic cosmic rays. There is probably a variety of mechanisms for injecting particles into the region of the radiation belts.

Properties of magnetic storms

Magnetic storms result from a change in the properties of the solar wind. As I have explained this is associated with enhanced solar activity such as the occurrence of giant flares. Typically the concentration of solar wind particles is increased to about $10^7 \, \mathrm{m}^{-3}$, instead of $5 \times 10^6 \mathrm{m}^{-3}$ in quiet times, and their velocity to about $900 \, \mathrm{kms}^{-1}$, instead of 300 to $500 \, \mathrm{kms}^{-1}$. The region of modified solar wind moves out from the Sun as a supersonic wave and when it reaches the magnetosphere it gives rise to several different phenomena, which include

(a)   a modification of the geomagnetic field, called a magnetic storm;
(b)   an increase in auroral activity at heights around 100 km;
(c)   a change in the ionosphere called an ionospheric storm.

When a solar flare preceding a magnetic storm is observed, the travel time of the particles from the Sun can be estimated by observing the time delay between the occurrence of the flare and the onset of the magnetic storm. The time is usually about 36 hours which is consistent with the solar wind velocity mentioned above.

The first effect of the enhanced solar wind is a change in the position of the magnetospheric boundary. As I mentioned earlier, the approximate position of this boundary is where the kinetic energy per unit volume in the solar wind ($\frac{1}{2}\rho v^2$) is equal to the magnetic energy density. This occurs at about $10 \, r_{\mathrm{E}}$; of course this is only a characteristic value of the distance on the solar side of the magnetospheric cavity. To obtain a simple but crude idea of what happens near the Earth's surface as a result of this, it is possible to study a model problem in which the magnetosphere boundary is taken as plane. We also idealise the structure of the Earth's magnetic field by assuming it to be that produced by a dipole at its centre with the dipole axis parallel to a line in the plane. Although this is not a true representation, it should be sufficient to provide an order of magnitude estimate of what happens.

Consider first the case of the quiet solar wind. All the field lines of the dipole field which should fill the space beyond the magnetosphere are confined within the boundary and the Earth. This confinement is produced by a current system at the boundary. There is a uniqueness theorem in magnetostatics which says that there is only one solution to any problem. In this case it is provided by placing an *image dipole* of the same strength and direction as the Earth's dipole at an equal distance on the opposite side of the boundary (fig. 58). This cancels out the Earth's field at the boundary but produces an additional field in the magnetospheric cavity. This leads to a field enhancement near the surface of the Earth, because there the image dipole produces a field in the same direction as the Earth's dipole. Because dipole fields fall off as $1/r^3$, the modification in the value of the field near the Earth's surface due to the image dipole is approximately $B_0(r_{\mathrm{E}}/2d)^3$, where B_0 is the strength of the magnetic field at the Earth's surface and d is the distance to the magnetospheric boundary. For the quiet solar wind and $B_0 \simeq 3 \times 10^{-5} \, \mathrm{T}$, $\Delta B \simeq 3 \times 10^{-9} \, \mathrm{T}$.

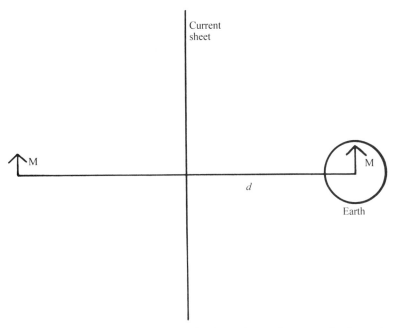

Fig. 58.   If the magnetospheric boundary is taken to be to a first approximation a plane, the field inside the cavity is identical to that produced by the Earth's dipole field and by the field due to an image dipole placed symmetrically with regard to the current sheet as shown.

If the solar wind is suddenly enhanced, the magnetospheric boundary is pushed inwards to about $6\,r_E$. A repeat of the above calculation then gives $3 \times 10^{-8}\,\text{T}$. This increase in the field occurs rapidly in a time which is comparable with the time that it takes the enhanced solar wind to move $4r_E$, when account is taken of the work done in compressing the magnetosphere. This is called the *sudden commencement* of the storm. Because the plane geometry is an idealisation, there is in reality also a spread of arrival time around the Earth's surface. The sudden arrival of the enhanced solar wind is followed by a more or less steady solar wind at the enhanced value for much longer than the time of sudden commencement. It might therefore be expected that the magnetic field would not change further during that time.

This is very far from what is actually observed. Shortly after the enhanced solar wind arrives, there is an increase in the number of trapped energetic particles in the radiation belts. There is some puzzle about these particles. They are not simply enhanced solar wind particles because their energies, up to keV electrons or 10 MeV protons, are very much higher than those of the solar wind particles (few eV electrons, few keV protons). Be this as it may, they exist and they also affect the strength of the magnetic field at the Earth's surface. Their main motion while they are trapped is one of movement up and down between the vicinity of the two magnetic poles of the Earth. But this is not their entire motion.

The ring current

The particles are moving in a magnetic field which varies in space and which has curved field lines. It is shown in Appendix 9 that both these properties of the magnetic field lead to drift motions of the particles. The drifts are round the Earth in a direction roughly perpendicular to the mirroring motion. The drift velocities are very much smaller than the velocities between mirrors, but they are in different directions for electrons and positive ions. As a result the particles in the radiation belts produce a *ring current*. This current produces a magnetic field at the Earth's surface which reduces the field generated in the Earth's interior. When the magnetic storm leads to an increased population of particles in the radiation belts, the larger ring current produces a reduction in the surface magnetic field, which is larger than the increase produced by the sudden commencement. Thus the *main phase* of the magnetic storm is one of reduced magnetic field. The main phase starts a few hours after the sudden commencement and the recovery to pre-storm conditions takes several days.

Superimposed on these two effects there are sometimes some further field excursions, which are associated with some additional currents which flow in the polar ionosphere. These are called polar substorms, but I will not discuss them any further. An idealised diagram of the change of the Earth's magnetic field during a magnetic storm has been shown in fig. 54. Some of the particles which were mentioned in the last paragraph can move down the field lines towards the magnetic poles of the Earth and this in turn can account for the great increase in auroral activity which accompanies magnetic storms. Magnetic storms can lead to a disruption of radio communication on Earth and that is one reason why there is interest in understanding the connection between events on the Sun's surface and at the Earth, as it would be useful to have warning of the disruption.

Aurorae

I have mentioned aurorae without saying precisely what they are. Most auroral displays occur at high latitudes; the *aurora borealis* in the northern hemisphere and the *aurora australis* in the southern hemisphere. The aurora is an optical display, which is concentrated in the atmosphere primarily in the polar regions although it is occasionally observable in southern England at times of great magnetic activity on the Sun. It is now known that it is produced by electrons moving along magnetic field lines towards the magnetic poles. These electrons have sufficient energy to excite the neutral atoms in the atmosphere from their ground states. The atoms subsequently decay to their ground states with emission of radiation. This typically occurs at heights of about 100 to 120 km. Above that height there are not enough neutral atoms to be excited and lower down all the electrons have been used up. The aurora is not itself a plasma phenomenon but it is an indication of dramatic plasma events occurring along the field lines at great distances (~ 2 to $8\,r_E$), which is where the particles which are later precipitated are first stored in the terrestrial field.

A more detailed discussion of the aurora is too complicated for this book but I will just make a few more comments on observations. The aurorae are of two types, discrete and diffuse. The diffuse aurorae, as I have mentioned, are the ones which have a significant contribution from radiation belt particles. The discrete aurorae seem to be somewhat different. Rocket flights into the discrete aurorae showed that the electrons responsible for them come from a very narrow energy range of 1 to 10 keV and that they were moving very close to parallel to the magnetic field lines. Satellite observations at an altitude far above that at which the auroral displays occur have found evidence of electrons streaming downwards, while ions are simultaneously accelerated upwards. These results are an indication of an electric field parallel to the magnetic field. This will automatically lead to particles which are moving close to parallel to the field lines. It is very difficult to maintain such an electric field because the particle motions which it produces tend to destroy it. Whatever is the origin of this electric field, its effect is to produce an electric current which is aligned with the primary magnetic field, because the electrons and ions are moving in opposite directions, and that this current in turn produces a magnetic field which is perpendicular to the original field. The overall configuration is certainly very complicated.

The auroral spectrum is complex. A bright aurora appears green or red to the eye, these colours being due to oxygen emissions at 557.7 and 630 nm. There is also important emission from nitrogen. It took a long time for the auroral spectrum to be understood because these spectral lines are forbidden lines of the type which I have discussed earlier as being important in the solar corona. Once again atmospheric densities are sufficiently low that radiative transitions can occur before a low lying excited state of an atom is de-excited by a collision.

The magnetospheric boundary; the bow shock

I now move to a discussion of conditions further from the Earth's surface. The motion of the solar wind towards the Earth is at all times highly supersonic. It is known that if an obstacle is placed in a supersonic flow, what is known as a *shock wave* must form. The shock front is a transition in which ordered kinetic energy is changed into random kinetic energy; the gas behind a shock has been heated but it is moving with a much lower bulk velocity. If the Earth did not possess a magnetic field, there would only be the shock transition (fig. 59). Because of the magnetic field, the magnetospheric boundary, magnetopause, occurs nearer to the Earth than the bow shock. This is shown schematically in fig. 50. The magnetopause is what is known as a contact discontinuity between solar wind and terrestrial particles at least to a good approximation.

Shock waves in laboratory gases are well understood. Although it is usual to idealise a shock transition as if it were a discontinuity (fig. 60a), it does actually have a finite thickness (fig. 60b). The thickness is related to the mean free path between collisions of the gas particles, because it is these collisions which enable the kinetic energy of ordered motion to be converted into kinetic energy of random motion. The shock thickness is a

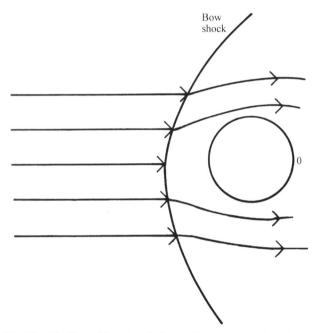

Fig. 59. The flow of the solar wind around a non-magnetised obstacle. There is a bow shock but no magnetopause.

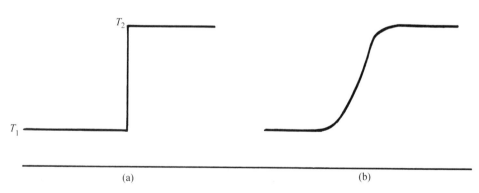

Fig. 60. A shock transition. (a) shows an idealised transition with a discontinuity in temperature from T_1 to T_2. (b) shows that in reality the transition has a width of a few collision mean free paths.

few mean free paths. If it is asked how thick is the bow shock transition in the solar wind, a surprising result is obtained. If estimates of collision frequency such as I made earlier in this chapter are used, it appears that the mean free path is comparable with the distance from the Sun to the Earth. In other words, there is insufficient space for a shock front.

In the 1950s scientists working on experiments aimed at controlled thermonuclear fusion in the laboratory were concerned with whether compression of a plasma could

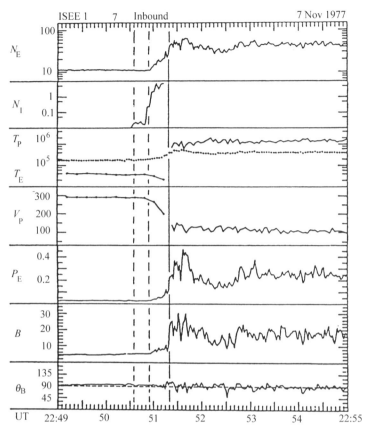

Fig. 61. The behaviour of many physical quantities at the bow shock as measured by instruments on the ISEE spacecraft (from Sckopke, N *et al.*, *Journal of Geophysical Research*, **88**, 6121, 1983).

produce a shock wave when the collision mean free path was greater than the size of the apparatus. If such collisionless shocks could exist, what was the process which enabled ordered energy to be converted into random energy in a distance short compared to the mean free path? It was thought that the magnetic field might solve the problem because particle gyration radii were much smaller than mean free paths and because it is difficult for two fluids to interpenetrate if they are both coupled to magnetic field lines. One of the major findings of early space experiments was that the bow shock did exist; it is what is known as a *collisionless shock*. The bow shock transition is found experimentally to be about 100 km thick.

The properties of the bow shock are quite different from those of shocks in laboratory gases. In these there is usually a smooth transition between the values of physical quantities on the two sides of the shock, as shown in fig. 60. The bow shock transition is much more irregular and it can vary significantly from one observation to the next. As I have explained already there is some ambiguity about whether they are measuring variations in space or variations in time or a mixture of both. In the case of shock transitions of the type shown in fig. 61, it is however clear both that there must be a shock and that the transition must be very irregular.

As I have discussed earlier, particularly in Appendix 6, a variety of waves can propagate in a plasma containing a magnetic field and it is believed that the role of simple particle/particle collisions in a collisional shock is replaced by the scattering of particles by waves, wave/particle interactions, in a collisionless shock. As these waves arise in a rather unpredictable manner the shock transition is very noisy. All the physical variables do, however, settle down to more or less steady values far enough on either side of the shock. A detailed understanding of the physics of the magnetospheric transition remains a major aim of present and future space missions.

The interplanetary magnetic field

I now move outside the magnetosphere to discuss the solar wind and the interplanetary medium a little further, because it is the impact of the solar wind and the interplanetary magnetic field which produces the magnetosphere. In Chapter 5 I explained that, once the solar wind has passed the radius where its kinetic energy density is equal to that of the solar magnetic field, its particles will move out into the interplanetary medium carrying the solar magnetic field with them. Within that radius the magnetic field is dominant and the solar wind is forced to corotate with the Sun leading to the magnetic braking which I mentioned earlier. Beyond the critical radius, the solar wind particles move in a straight line with constant velocity to a good approximation but, because the Sun is rotating and

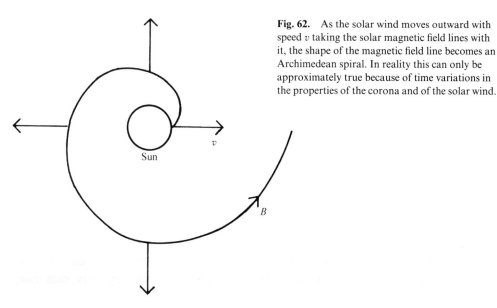

Fig. 62. As the solar wind moves outward with speed v taking the solar magnetic field lines with it, the shape of the magnetic field line becomes an Archimedean spiral. In reality this can only be approximately true because of time variations in the properties of the corona and of the solar wind.

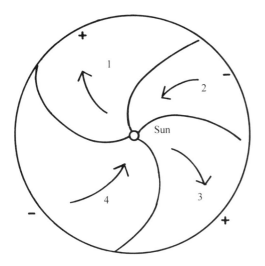

Fig. 63. The direction of the interplanetary magnetic field as measured by a spacecraft close to Earth (schematic). As the Sun rotates, the spacecraft samples the magnetic field on the circle shown. In regions 1 and 3 the field was mainly positive and in regions 2 and 4 mainly negative indicating the sector pattern of the solar magnetic field which is indicated. In region 2 the field changed sign during a magnetic storm and some negative fields were also measured in region 1 just before the storm.

with it the footprints of the magnetic field lines, the magnetic field takes up a spiral structure known as an Archimedean spiral (fig. 62).

If the magnetic field of the Sun were simply dipolar, this would produce an equally simple pattern in the ecliptic plane containing the Earth, once allowance was made for the tilt of the dipole axis to the plane of the ecliptic. We have however seen that the actual field at the solar surface is very complex, more so at solar maximum than at solar minimum, and that the solar wind is produced primarily in the cooler regions of the corona known as coronal holes. Figure 63 is a schematic picture of measurements of the direction of the interplanetary magnetic field made by a spacecraft. Although the spacecraft only samples a very small part of the interplanetary medium close to the Earth, the rotation of the Sun allows it to study (in one solar rotation period) the whole of the magnetic field structure in the ecliptic plane. The diagram shows two things. It shows the direction of the measured magnetic field and the inferred spiral structure of the interplanetary field. It can be seen that there is a sector structure, with the direction of the field alternating from one sector to the next. This regularity was briefly disturbed when there was a magnetic storm and a positive field replaced a negative field. This is not surprising because the magnetic storm solar wind carries the solar magnetic field to the Earth's vicinity much more rapidly than the steady solar wind. This discussion is believed to be only a first approximation to the truth, but any more detailed discussion is beyond the scope of this book.

The interplanetary medium

The interplanetary medium, both within the Earth's radius and out to and beyond the outer planets, is not empty. It contains not only the smaller objects in the solar system, asteroids, comets, meteors, meteorites, but also a general interplanetary dust and interplanetary plasma. It is with the latter that I am concerned in this paragraph.

Although most detailed information about the interplanetary medium is now obtained by use of spacecraft, it is still possible to obtain some valuable knowledge from ground-based observations. In the 1960s investigation of the interplanetary medium led to a major astronomical discovery. In Appendix 8 I discuss the propagation of electromagnetic waves in a uniform plasma, showing how their phase and group velocities are both different from c. If the properties of the plasma, such as its density, are irregular in space and time, the waves can be scattered by these irregularities. This effect is likely to be important in the radio region of the spectrum. If a distant radio source is viewed through a fluctuating plasma, its apparent position will vary as a result of the scattering of its radiation. This process is known as *scintillation*. It is similar to the twinkling of stars produced by the scattering of light in the Earth's atmosphere.

In the 1960s, A Hewish at Cambridge constructed a radio telescope array to study the scintillation of radio sources, with the aim of determining the characteristic lengths and times of fluctuations in the interplanetary medium. He was successful in doing this but the telescope made another discovery of great significance. In order to study the plasma fluctuations, the instrument required a time resolution of less than a second. At that time radio telescopes tended to work with much longer integration times (corresponding to the exposure time of an optical telescope) because it was not thought likely that astronomical sources of radio waves would vary with a very short timescale. It was therefore a total surprise when Hewish's student Jocelyn Bell discovered a new type of radio source, *pulsars*, which emitted pulsed radiation with a period of order 1 s. It was soon realised that this was the first definite discovery of *neutron stars*. Now pulsars are known with periods as short as ms. Pulsars have also enabled information to be obtained about the properties of the interstellar plasma but that is outside the scope of this book. The aim of including this paragraph is to demonstrate the interrelation between different branches of astronomy.

The magnetospheric tail

The solar wind forms the magnetospheric cavity. It is easy to understand, at least in principle, what happens at the near side to the Sun but the magnetosphere must also be closed at the far side. This implies that from a large distance the solar wind would appear to flow past an obstacle containing the Earth but very much larger than the Earth. Planetary magnetospheres have long tails pointing away from the Sun. The length of the Earth's tail is $\sim 200 \, r_E$, which has been detected by lunar orbiters. This over an order of magnitude larger than the $\sim 10 r_E$ distance on the near side. Jupiter's is much longer still, due to its comparatively very large intrinsic magnetic moment, and it reaches part of the Saturnian orbit.

As mentioned earlier the magnetosphere is separated from the solar wind by a surface called the magnetopause, across which the solar wind pressure is balanced by the total internal magnetic and particle pressure. The simplest model for the magnetospheric

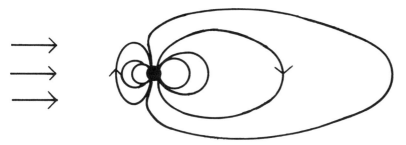

Fig. 64. A closed magnetosphere in which the Earth's magnetic field is confined in the cavity. It is approximately dipolar at low latitudes but is modified at high latitudes.

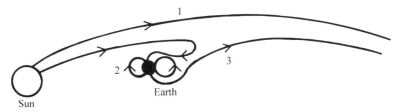

Fig. 65. An open model of the magnetosphere showing three types of field lines: 1 a field line pulled out from the Sun by the solar wind, 2 a dipole field line of the Earth and 3 a field line which connects the Sun, Earth and interplanetary space.

transition is one in which the interplanetary magnetic field is totally isolated from the magnetospheric magnetic field and in which no solar wind particles cross the magnetopause, except possibly as a result of an instability (fig. 64). A second possibility is that of an open magnetosphere in which there are some magnetic field lines which have one end anchored on the planetary surface and the other on the Sun (fig. 65). This can only happen if there is reconnection of magnetic field lines of the type which I mentioned in Chapter 4 and when I discussed solar flares. The open/closed debate has been a hot topic for some decades, but evidence is now accumulating in favour of the open model first suggested by J W Dungey in 1961.

In this picture the formation and structure of the magnetosphere is dynamic. Magnetic field lines from the Sun are being pulled out continually to and past the Earth by the solar wind particles. It is believed that the observed magnetospheric structure, and in particular its boundary on the side of the Earth furthest from the Sun, depends on a continuous process of magnetic field reconnection. The solar magnetic field lines are supposed to reconnect with terrestrial magnetic field lines leading as a result to a partially open magnetosphere. The process of reconnection is illustrated again in fig. 66 and the way in which the magnetic field structure is thought to evolve is shown in fig. 67. In this fig. the arrows indicate the direction of the solar wind plasma, and of other motions, which can carry the magnetic field lines around and drive reconnection. The essential process is the creation of merged magnetic field lines, which have one end

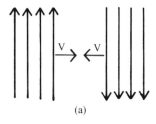

(a)

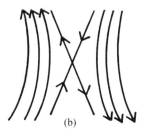

(b)

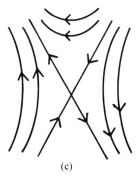

(c)

Fig. 66. A process of steady state reconnection, showing successive field lines approaching and reconnecting.

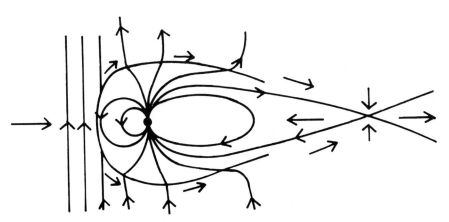

Fig. 67. The evolution of merged solar and terrestrial field lines being swept back to form the magnetotail. The arrows not on field lines show the direction of plasma flow. Reconnection is occurring in the tail.

attached to the Earth (or to any other planet with a magnetosphere) and the other in interplanetary space. Eventually the size of the magnetosphere is limited by reconnection at the magnetotail. If the open model of the magnetosphere is correct, it is obviously possible for some solar particles to get directly into the magnetosphere and hence into the radiation belts and the aurorae.

Solar wind composition and velocity distribution

In the simplest discussion of the solar wind, it is assumed that it consists of a mixture of electrons and protons and that they all move with a single velocity. This is incorrect in two respects. Although hydrogen is the most abundant element in the solar atmosphere, there is about one helium atom for every ten hydrogen atoms and in reality the solar wind contains a significant proportion of ^{4}He nuclei (α-particles). In fact, the relative number of protons and α-particles varies with time; sometimes there are fewer α-particles than would be expected from the surface composition of the Sun and sometimes more. This must presumably tell us something about the process by which the solar wind is produced or about abundance irregularities in the outer solar atmosphere. The variation in composition means that studies of the solar wind cannot provide a reliable value for the surface abundance of solar helium.

The second point is that the particles have a distribution of velocities. Crudely speaking this can be described as a temperature, but there is only really a temperature if the distribution of velocities is Maxwellian (see Appendix 1). To have a Maxwellian distribution it is necessary to have collisions. Although the collision frequency is high enough close to the solar surface, I have explained earlier that the mean free path in the interplanetary plasma is of order 1 AU. Observations show that the velocity distribution is non-Maxwellian particularly in the sense that the spread of velocity along the magnetic field is greater than that across the field. In terms of some crude measure of temperature, $T_\parallel > T_\perp$. In any case the pressure of the solar wind plasma is not isotropic, $P_\parallel > P_\perp$.

Planetary magnetospheres

Although this chapter is concerned with solar–terrestrial relations, I finish with a few words about other planets. It is possible to distinguish three types of bodies as far as their interactions with the solar wind are concerned. Those with their own magnetic fields of internal origin – Mercury, Earth, Jupiter, Saturn, Uranus, Neptune – interact with the magnetised solar wind which more or less confines the planetary magnetic fields to magnetospheres. Even if the solar wind particles move with constant velocity, their volume density drops off with distance from the Sun, which means that even if all planets had similar magnetic fields those far from the Sun would have larger magnetospheres. Unmagnetised bodies, which have atmospheric particles which can form an ionosphere (Mars, Venus, comets), also interact with the solar wind producing currents and creating a magnetic cavity. This is known as an induced magnetosphere. The Moon is more like an insulator, non-magnetic and non-conducting, and it absorbs and neutralises the solar wind particles giving a cavity and wake downstream.

The most spectacular plasma phenomena in the planetary system occur in the magnetosphere of Jupiter. Radio emission from Jupiter was detected in the 1950s and it was soon realised that the higher frequency radiation must be produced by energetic particles spiralling in a magnetic field (synchrotron radiation). In effect jovian radiation belts were discovered before the terrestrial van Allen belts. The jovian magnetosphere has since been studied *in situ* by four planetary probes (Pioneer 10, 1973, Pioneer 11, 1974, Voyager 1 and Voyager 2, 1979) so that many local measurements have been made. Lower frequency radiation (5 to 40 MHz) is closely related to the presence of the satellite Io which is inside the magnetosphere at a distance of 6 jovian radii. Io, which is highly volcanic, continuously loses gas, which is then ionised and forms a plasma torus around Io. There is a lack of symmetry in the Jupiter, Io, plasma torus configuration and this leads to generation of plasma waves, the acceleration of electrons and the emission of the low frequency radiation. The whole interaction between Jupiter, Io and the magnetosphere is complex and not fully understood.

The heliosphere boundary

The solar wind in effect pushes the boundary of the Sun away from the Sun's surface. So far I have discussed the influence of the solar wind on the surroundings of the Earth and other planets. However, as well as confining planetary magnetic fields into magnetospheres, the solar wind is producing a cavity surrounding the Sun in the local interstellar medium. This cavity is known as the *heliosphere*. When I discussed the corona and solar wind in chapter 5, I explained that a static corona could not exist in pressure balance with the interstellar medium. Once the solar wind is introduced there will be a surface surrounding the Sun at which the solar wind pressure balances the pressure of the interstellar gas and there will be a heliospheric boundary whose structure is basically similar to that of the magnetosphere. In some ways the heliospheric boundary can be considered to be the effective surface of the Sun.

Obviously the position of the boundary depends both on the strength of the solar wind and on the properties of the local interstellar medium. The components of the interstellar medium include the ionized and neutral gas, the magnetic field and cosmic rays. The gas, magnetic field and cosmic rays are known to exert pressures which are of comparable magnitude. The position of the boundary is not known with any great accuracy but it is probably of order 100 times the Earth's distance from the Sun. The heliosphere thus contains most of the solar system but not the most distant comets, known as the Oort cloud. As I have mentioned in Chapter 5, recent observations made by the *Ulysses* spacecraft have shown that, at least during solar minimum, the solar wind from the poles has a higher speed than the speed in the ecliptic. This should lead to a heliospheric boundary which is further from the Sun in the polar direction.

Summary of Chapter 7

Although the main influence of the Sun on the Earth is through its emissions in the optical and neighbouring regions of the electromagnetic spectrum, there is an additional effect produced by the corpuscular emission from the Sun, the solar wind. The solar wind particles confine the Earth's magnetic field in what is called the magnetospheric cavity. The boundary of the cavity is about ten Earth radii from the Earth on the sunward side but more than ten times this distance in the magnetotail far from the Sun. A study of the magnetosphere is of interest for two distinct reasons. There is naturally interest in trying to obtain a full understanding of the manner in which the Sun influences the Earth, particularly as fluctuations in the strength of the solar wind lead to magnetic storms which interfere with radio communication on Earth. In addition the magnetosphere provides a natural laboratory in which the physics of ionised gases or plasmas can be studied in conditions which cannot be achieved on Earth. As an example laboratory experiments had been unable to determine whether or not collisionless shock waves could exist when spacecraft measurements showed that the magnetospheric bow shock was a collisionless shock.

Spacecraft are able to make a wide variety of measurements of the properties of space plasmas and magnetic fields. They collect such a large quantity of data that there are major problems in transmitting it to Earth and analysing it afterwards. Care must be taken in the design of experiments so that they measure the true properties of the space plasma rather than the properties as modified by the presence of the spacecraft. In addition, because of the motion of the spacecraft, it is not always easy to distinguish between spatial and temporal changes in the property of the medium. The CLUSTER mission being launched in 1996 consists of several spacecraft whose spatial configuration can be varied and this will help to resolve these problems.

Magnetic storms were studied before their origin was known. It is now known that they occur when enhanced solar activity increases the strength of the solar wind. This in turn compresses the magnetosphere and increases the strength of the magnetic field at the Earth's surface. The first important discovery of spacecraft other than the solar wind was the van Allen radiation belts. Charged particles in the magnetosphere are observed to be trapped in the Earth's magnetic field, by what is known as the magnetic mirror effect. At the time of a magnetic storm, the number of trapped particles increases and their drift motions produce a ring current which reduces the magnetic field at the Earth's surface, more than compensating for the increase in the sudden commencement of the storm. Particles moving out of the radiation belts towards the Earth's surface can interact with the neutral atmosphere to produce the brilliant auroral displays observed in polar latitudes.

Because the motion of the solar wind and the associated magnetic field is continuous, the structure of the magnetosphere must be dynamic. Reconnection of solar and terrestrial field lines is believed to occur, leading to the injection of some solar wind particles into the magnetosphere. Because of the rotation of the Sun, which carries the footpoints of a solar field, the field has a spiral structure in interplanetary space. The solar wind interacts with the magnetic fields of other planets in the solar system. In particular Jupiter has an extremely large magnetosphere and the most intense plasma phenomena in the solar system involve interactions between the solar wind, Jupiter and its satellite Io.

Eventually the influence of the solar wind must end when its pressure is balanced by the pressure of the local interstellar medium. The region in which the solar wind is important is known as the heliosphere. All the major components of the solar system are situated well within the heliosphere.

8 The origin of the solar system

Star formation

The subject of star formation is not very well understood. It is not difficult to understand why this could be the case. The formation of a star like the Sun is believed to occupy a period of order 10^7 yr. This is both a long time and a short time. It is long in the sense that observations can only give us a snapshot of the process of star formation. If sufficient forming stars can be observed, it might be possible to deduce a plausible evolutionary sequence, but this is not the same as direct observation of evolution. The time of formation is a short time in the sense that it is only about one thousandth of the main sequence lifetime of the star. In our Galaxy the rate of star formation has probably declined slowly with time and this means that the number of stars currently forming is probably less than one thousandth of the total number of stars. This means that forming stars are rare. There is one point which counters this slightly. Star formation is observed to be an erratic process which means that at any time there are regions in the Galaxy where there is no significant star formation and others which are very active sites of star formation. Those active sites of star formation which are nearest to the Sun are the focus of most observational attention.

A theoretical study of star formation is very difficult. A star must be formed out of interstellar gas but the typical internal density of a star is of order 10^{24} times that of the average interstellar gas. Even if we start with a relatively dense gas cloud, an increase of density of order 10^{20} is required. The interstellar gas contains a magnetic field and also has rotational energy. Both of these effects resist the tendency of gravity to cause clouds to contract to form stars. In addition they ensure that any contracting cloud of gas is unlikely to be spherical. Such a departure from sphericity makes numerical calculations much more difficult. In addition it is easy to show that any cloud which becomes spontaneously unstable to gravitational collapse is likely to be of star cluster mass rather than of stellar mass. This means that individual stars are likely to emerge out of the successive fragmentation of a massive cloud. A realistic theoretical study of the star formation process is very difficult even with large modern computers.

For a long time there was also not very much observational information about star formation. This was not only because forming stars are rare but also because the cold dense gas clouds in which star formation starts are highly opaque to optical radiation. It has been the developments in infrared and millimetre wave astronomy which have provided a wealth of information about star-forming regions in our own Galaxy as well as in other galaxies. Of particular importance were observations with the Infrared Astronomy Satellite (IRAS) which was operational for 9 months in 1986. The various observations of star-forming regions have shown that star formation is a very complex process. The original simple picture of a spherical cloud contracting to form a spherical star has been replaced by a formation process which can include both accretion from a disk and a bipolar outflow in a direction perpendicular to the disk (fig. 68). It has also become clear that star formation is a rather inefficient process. Much of the material, which is in clouds which become star-forming regions, is returned to the general interstellar medium rather than being incorporated in stars.

   The problem of star formation is one of the most important problems in astronomy because it influences our observations of the most distant objects in the Universe. The most distant galaxies are observed in the light of their stars and to understand their properties we shall need to understand something about how star formation occurred in them and whether it is similar to or different from star formation in our Galaxy today. A question of much more local interest is the ultimate end point of star formation. More than half of the known stars in the solar neighbourhood are members of binary or multiple systems of stars; the Sun in contrast possesses a planetary system. Do all stars

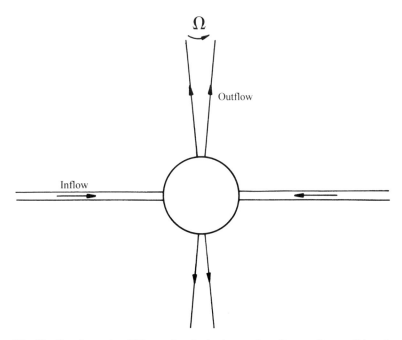

Fig. 68.   Star formation. This may involve both accretion of matter from a disk and outflow of matter along the axis of rotation.

fall into one of these two classes or are there stars which have always been genuinely single stars?

Formation of the solar system

It is therefore of interest to ask what observations of the present Sun and of the solar system might be able to tell us about the origin of the system and thence of star formation in general. Ever since scientists started serious speculations concerning the formation of the solar system, there have been two competing types of theory. The first now goes by the name of the *solar nebula theory* and it was originally proposed independently by Kant and by Laplace. They suggested that the solar system was originally a flattened rotating cloud of gas out of which the cental regions formed the Sun and the outer regions formed the planets and other lesser objects. This is overwhelmingly the most popular theory today. If it proves to be a valid explanation for the solar system, it is plausible that all single stars, or at least those whose overall properties are not too different from the Sun should have planetary systems. This is obviously significant in any discussion of the probable evolution of life elsewhere in the Universe.

The second general idea is one of interaction. This was originally proposed by Buffon, taken up by Chamberlain and Moulton and developed by Jeans and by Jeffreys, so that by the 1930s it was *the* theory of the origin of the solar system. According to this theory the close passage of another star pulled a filament of material out of the Sun and this material subsequently went into orbit around the Sun and formed the planets. If such a theory were correct, planetary systems might be very rare because close enough stellar passages would be very infrequent. In fact that was not quite so obvious when Jeans popularised the theory because he believed that the Universe was 10^{12} yr old rather than 10^{10} yr, as will be mentioned later in the chapter. In addition, there would be no reason why the Sun and the solar system should have essentially the same age. There is now a general belief that the interaction theory is not only inherently improbable (not that this really matters while we only have clear knowledge of the existence of one planetary system) but more important that it probably would not work.

A variant of the interaction model has been proposed by M M Woolfson, although he has not succeeded in challenging the general support given to the solar nebula theory. One of the major problems in the original interaction theory is the amount of energy that has to be provided to remove matter from the outer layers of the Sun. Woolfson suggests instead that the Sun formed in a cluster or association of stars and that the planetary system was formed when the Sun captured some of the outer layers of a protostar. The amount of energy required to pull material out of a protostar, which has a very much larger radius that a main sequence star, is much less than in the normal interaction theory. In addition stars in a cluster can be much closer together than in the general field, which increases the probability of an interaction. However it remains an unlikely process. There is no objection to the idea that the Sun was formed in a cluster provided that the cluster's properties were such that it dispersed when a supernova exploded in it.

Table 5.   The radioactive decay chain of ^{238}U

$$^{238}_{92}U \xrightarrow[\alpha]{4.5 \times 10^9 \text{ yr}} {}^{234}_{90}Th \xrightarrow[\beta]{24.1 \text{ d}} {}^{234}_{91}Pa \xrightarrow[\beta]{6.7 \text{ hr}} {}^{234}_{92}U \xrightarrow[\alpha]{2.5 \times 10^5 \text{ yr}} {}^{230}_{90}Th$$

$$\xrightarrow[\alpha]{80 \text{ yr}} {}^{226}_{88}Ra \xrightarrow[\alpha]{1.6 \times 10^3 \text{ yr}} {}^{222}_{86}Rn \xrightarrow[\alpha]{3.8 \text{ d}} {}^{218}_{84}Po \xrightarrow[\alpha]{3.1 \text{ m}} {}^{214}_{82}Pb \xrightarrow[\beta]{26.8 \text{ m}} {}^{214}_{83}Bi$$

$$\xrightarrow[\beta]{19.7 \text{ m}} {}^{214}_{84}Po \xrightarrow[\alpha]{164 \text{ }\mu s} {}^{210}_{82}Pb \xrightarrow[\beta]{21 \text{ yr}} {}^{210}_{83}Bi \xrightarrow[\beta]{5.0 \text{ d}} {}^{210}_{84}Po \xrightarrow[\alpha]{138 \text{ d}} {}^{206}_{82}Pb$$

The half-life of each unstable isotope is shown together with the decay mechanism (α-decay or β-decay).

I shall return to these theoretical ideas relating to the origin of the solar system when I have discussed what observational evidence we have that bears on its origin.

The age of the solar system

As should be clear from what I have said earlier, there is only one way by which we can obtain direct evidence concerning the age of the Sun. This comes from a discussion of the ages of objects in the solar system, specifically lunar and terrestrial rocks and meteorites. There is a very strong presumption that the Sun must have existed throughout the past history of the solar system but, as I have already mentioned, it could in principle be older. Ages of solar system rocks are determined by a study of the long-lived radioactive elements which they contain. There are several elements and their decay products which can be used but I shall concentrate on the use of the isotopes of uranium, ^{235}U and ^{238}U, both of which decay to isotopes of lead. Their respective half-lives are 0.7×10^9 yr and 4.5×10^9 yr. The radioactive decay chain for ^{238}U is shown in Table 5. The first decay takes so much longer than all the subsequent ones that to a first approximation it can be assumed that an atom is either ^{238}U or ^{206}Pb. The same is true of ^{235}U and its decay product ^{207}Pb.

It is assumed that the solar system started with a certain abundance of radioactive isotopes and that these have freely decayed since that time. It can be assumed that they have not been added to because, although radioactive elements continued to be produced in stars and to be expelled into the interstellar medium, the mean density of the solar system is so much higher than the average density of the interstellar medium that any fractional enrichment is now negligible. If solid objects form in the solar system and are never subsequently melted, there will be a gradual and predictable conversion of U into Pb and enrichment of ^{206}Pb and ^{207}Pb relative to an isotope of lead, such as ^{204}Pb, which does not result from radioactive decay. In different solid objects formed at the same time there may be a variation of the ratio of total uranium to total lead content, because of chemical processes associated with the formation of the solids. However the relative isotopic abundances of U and of Pb should not be affected by such chemical processes. This leads to a method of dating the formation of chemically different rock samples of the same age, which I now describe.

The ages of rock samples

Suppose that in a unit mass of rock at the time of its solidification there are $N_0(A)$ atoms of a radioactive isotope A, which has a decay rate λ. Then the number of atoms of isotope A changes with time according to the equation

$$\frac{dN(A)}{dt} = -\lambda N(A). \tag{8.1}$$

This equation integrates immediately to give

$$N(A) = N_0(A)\exp(-\lambda t), \tag{8.2}$$

where I am taking the zero of time to be the instant of solidification. If the decay product, isotope B, started with initial abundance $N_0(B)$, its abundance at time t would be

$$N(B) = N_0(B) + N_0(A)(1 - \exp(-\lambda t))$$
$$= N_0(B) + N(A)(\exp \lambda t - 1) \tag{8.3}$$

where B is assumed stable.

I can now apply the above equation to the decay of the two isotopes of uranium to lead. Thus

$$N(^{206}\text{Pb}) = N_0(^{206}\text{Pb}) + N(^{238}\text{U})(\exp \lambda_{238} t - 1),$$
$$N(^{207}\text{Pb}) = N_0(^{207}\text{Pb}) + N(^{235}\text{U})(\exp \lambda_{235} t - 1), \tag{8.4}$$

The two equations (8.4) can be combined to give

$$\frac{N(^{207}\text{Pb}) - N_0(^{207}\text{Pb})}{N(^{206}\text{Pb}) - N_0(^{206}\text{Pb})} = \frac{N(^{235}\text{U})}{N(^{238}\text{U})} \frac{(\exp \lambda_{235} t - 1)}{(\exp \lambda_{238} t - 1)} \equiv f(t). \tag{8.5}$$

In the expression which I have called $f(t)$, $N(^{235}\text{U})/N(^{238}\text{U})$ is the isotopic ratio of uranium at the present time, which is always found to be about 0.0072. This explains why I can write $f(t)$ as a function of time alone. It is also crucial to the whole method of radioactive dating.

It is now convenient to express all of the lead abundances in terms of the abundance of an isotope, ^{204}Pb, which is not a radioactive decay product. I first rewrite equation (8.5)

$$N(^{207}\text{Pb}) - N_0(^{207}\text{Pb}) = f(t)[N(^{206}\text{Pb}) - N_0(^{206}\text{Pb})]. \tag{8.6}$$

I then divide (8.6) by $N(^{204}\text{Pb})$, which could equally be written $N_0(^{204}\text{Pb})$, to give

$$\frac{N(^{207}\text{Pb})}{N(^{204}\text{Pb})} - \frac{N_0(^{207}\text{Pb})}{N(^{204}\text{Pb})} = f(t)\left[\frac{N(^{206}\text{Pb})}{N(^{204}\text{Pb})} - \frac{N_0(^{206}\text{Pb})}{N(^{204}\text{Pb})}\right]. \tag{8.7}$$

If it is now supposed that I have a set of rock samples of the same age, they may differ in their original relative abundances of U and Pb, because of processes separating the two elements during the rock formation, but they should all have had the same relative isotopic distribution at that time. Thus $N_0(^{207}\text{Pb})/N(^{204}\text{Pb})$ and $N_0(^{206}\text{Pb})/N(^{204}\text{Pb})$ should be the same for all of them. If that is the case equation (8.7) can be schematically written

$$y = mx + k, \tag{8.8}$$

the equation of a straight line, where

$$y = N(^{207}\text{Pb})/N(^{204}\text{Pb}), \quad x = N(^{206}\text{Pb})/N(^{204}\text{Pb}) \text{ and } m = f(t), \tag{8.9}$$

and k is a constant having the same value for all of the samples. If the observed values of x and y for a variety of samples are plotted and the points do lie on a straight line, the value of $f(t)$ can be determined and hence the ages of the rocks are known.

This method can obviously be applied to a rock stratum on Earth, which results from one period of mountain building. Provided that it has a differentiated chemical composition, different samples will have different lead isotopic abundances and the method outlined above can be used. The method has also been used to study meteorites with the assumption that they are solid bodies formed very early in the history of the solar system and that their ages are essentially the same. In this case the justification for the assumption is that the method works. Results obtained by a study of three different types of meteorite and of deep terrestrial oceanic sediments are shown in fig. 69. The data represent a wide range of values of x and y and they do lie close to a straight line appropriate to an age of 4.55×10^9 yr. The deep oceanic sediments are supposed to represent that part of the Earth's crust which has been least affected by continuing

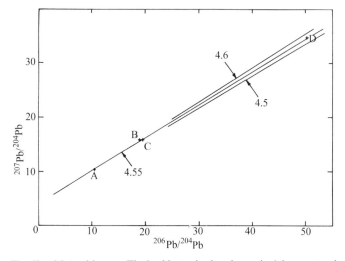

Fig. 69. Meteoritic ages. The lead isotopic abundances in A iron meteorites, B oceanic sediments, C chondritic meteorites, and D an achondrite lie on a line which indicates a common age of 4.55×10^9 yr. The lines for 4.5 and 4.6×10^9 yr are also shown.

geological activity on the Earth. It is found that terrestrial rocks have a wide variety of ages, which do represent continuing tectonic activity, but that none of the rocks is as old as the deep oceanic sediments. The lunar rocks, that have been studied, are typically very much older than the terrestrial rocks, but there does appear to be a spread in ages indicating that the Moon did have an extensive period of seismological history. The oldest rocks discovered have an age approaching 4.5×10^9 yr.

The original technique used by Rutherford to date the formation of terrestrial rocks was simpler in principle to what I have described above but also less accurate in practice. The decay of uranium to lead produces helium; 8 α-particles in the decay of each ^{238}U and 7 α-particles in the decay of each ^{235}U. Rutherford argued that, if he could measure the amount of He found together with U, he could decide how much uranium had decayed and hence determine the age of the rock. It is not difficult to obtain the equation

$$N(^4\text{He}) = 8N(^{238}\text{U})(\exp \lambda_{238} t - 1) + 7N(^{235}\text{U})(\exp \lambda_{235} t - 1). \tag{8.10}$$

In equation (8.10) $N(^4\text{He})$, $N(^{238}\text{U})$ and $N(^{235}\text{U})$ can be measured and λ_{238} and λ_{235} are known, so that t can be determined. Although the above method is simple in principle (and it can be generalised to the case in which ^{232}Th and other α-emitting radioactive isotopes are present), it suffers from two problems arising from the fact that helium is a gas. The first is that, if the rock is at all porous, some of the helium may migrate from the region where it is produced. The second is that, when the rock is broken open for analysis, it is difficult to ensure that all the helium is collected. For both of these reasons, it is likely that less than the full amount of helium will be measured and this will lead to an underestimate of the age of the rock.

Rutherford's announcement that there were terrestrial rocks with ages greater than a few hundred million years was sensational. At that time it was thought that the Sun was radiating energy which was released by its slow contraction. If that were the case its past life can have taken only a time comparable with the thermal or gravitational timescale discussed in Chapters 3 and 5,

$$t_{\text{Th}} = GM_s^2/L_s r_s. \tag{3.14}$$

For the Sun this allows a past life of tens of millions of years but not hundreds of millions of years. It was difficult to see how the Earth could be older than the Sun but there was no known source of energy which could have provided the Sun's luminosity for hundreds of millions of years. The existence of the atomic nucleus was not at that time suspected so that there was no idea of the possibility of nuclear transmutation with release of energy. Chemical energy was even less effective than gravitational energy. In the period that followed until nuclear energy was discovered, one suggestion was that of Jeans that in the centres of stars there could be a complete annihilation of mass. If that were possible, the potential energy supply would be more than a hundred times what can be obtained from nuclear fusion reactions. This led Jeans to believe that stars could live for a very long timescale of order 10^{12} yr. This in turn could allow more time for the interaction process which he believed led to the formation of the solar system. The long-timescale suggested by Jeans could not survive Hubble's discovery of the expansion of the Universe, with its associated timescale, and the increased knowledge of the properties of atomic nuclei. In particular matter annihilation was not possible, while antimatter, which could annihilate with matter, was not found to be present.

I now return to the general question of the age of the solar system. The results described above suggest that the Earth, Moon and meteorites reached something like their present form about 4.55×10^9 yr ago. If the solar system did originate as one entity, this should also be the approximate age of the entire solar system. Is there any further evidence bearing on this point? The presence of fossils in rocks indicates that the Earth was a suitable place for life at the time that the rocks were laid down. This implies that the Sun was luminous at that time. The oldest known microscopic fossils are about 3 $\times 10^9$ yr old which can be taken as evidence that the Earth and the Sun were associated and that the Sun had a high luminosity that long ago. It seems inconceivable that they have not been associated for the whole 4.55×10^9 yr, but it does not appear possible to prove directly that the Sun is not a lot older than that. Here we can only rely on theoretical ideas.

Chemical evidence for the origin of the solar system

If the whole solar system originated together at the same time, there should be a similarity in the chemical compositions of different parts of it. I therefore next turn to the evidence on this point. It is possible to study the chemical composition of different objects in the solar system as follows:

(a)   Direct chemical analysis; the Earth (atmosphere, oceans and crust), Moon rocks (gathered by Apollo missions), meteorites (gathered on Earth);
(b)   Spectroscopic observations; outer-layers of the Sun, planetary atmospheres, gaseous regions of comets.

In addition some clues to possible chemical composition are provided by the mean densities of planets and satellites, by the existence of planetary magnetic fields (which are produced by currents in conducting media), by seismic studies of the Earth and the Moon and by the surface appearance of planets and satellites. I have already discussed the chemical composition of the Sun in Chapter 2 and I shall not repeat that discussion here.

First I should make a comment on the possible significance of any results which might be obtained. The Sun and the solar system, whether formed together or not, have originated from the interstellar medium in the Galaxy. The chemical composition of the interstellar medium varies with time as a result of nuclear reactions in stars followed by mass loss from stars. At any given time the interstellar medium in any locality will have a composition which will be the original composition of any objects born there. The composition will however vary both in space and time. If the whole solar system originated in one event, all of its components should have essentially the same composition, both elemental and isotopic, when allowance is made for any segregation effects which occurred during the formation phase and subsequent history. This similarity of chemical composition is a necessary condition for common origin but it is not a sufficient condition. Theories of galactic evolution sugest that the chemical composition of the interstellar medium does not change much in 10^9 yr. It is therefore unlikely that a difference in age of that amount could be ruled out from observations of essentially similar compositions for different parts of the solar system.

The composition of the Earth

In the case of the Earth it is only possible to make direct observations of the atmosphere, oceans and crust and these represent only 0.5 per cent of the mass of the Earth. About 68 per cent of the Earth's mass is in the mantle, which is an elastic solid of higher mean density than the crust, and the remainder is in the core, which is largely liquid and which is of higher density than the mantle. There is a small solid inner core. Figure 70 is a schematic picture of the structure of the Earth. The density distribution in the Earth and its division into solid and liquid phases has been deduced by a study of the propagation of earthquake waves. The speed of the waves depends on the pressure/density relationship in the Earth. There are two types of wave, longitudinal and transverse, but only longitudinal waves can propagate in the liquid core. When waves from a single earthquake are observed at many seismological stations, their travel times provide

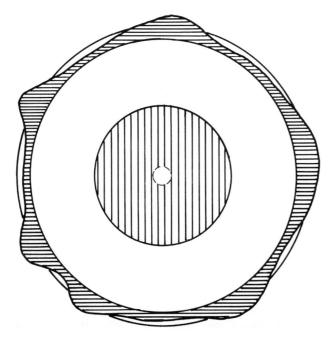

Fig. 70. The structure of the Earth. The vertically striped region is the liquid core and a small inner solid core is indicated. Outside the core is the mantle. The horizontally striped region is the crust above which are the oceans. (The diagram is not to scale.)

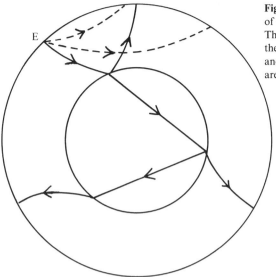

Fig. 71. Earthquake waves. The propagation of waves from an earthquake at E is shown. The dashed lines denote waves not entering the core. For the solid lines, several reflections and refractions at the core–mantle interface are shown.

much information about the structure of the Earth's interior. The propagation of earthquake waves is illustrated in fig. 71.

The Earth's magnetic field is believed to arise from electrical currents flowing in the Earth's interior. This can only occur in the liquid core, which must therefore be electrically conducting. As I have explained in Chapter 4, magnetic fields decay if they are not regenerated, and the magnetic field of the Earth would decay in a very small fraction of the present age of the Earth. This means that a dynamo process must currently be acting in the Earth. The core is presumably liquid because it is hot. One source of heating is the decay of radioactive elements. There will have been more radioactivity when the Earth was younger and much of the Earth may then have been molten. Conditions could then have represented those in a blast furnace when some elements are observed to fall in the gravitational field and others to rise to the surface. It is consistent with this picture that metals such as iron and nickel should have concentrated to produce a conducting core and that the outer layers should primarily contain silicates. When all of these considerations about the Earth's interior are taken into account an estimate can be made of the overall composition of the Earth.

Meteoritic composition

Only very limited samples of the Moon have been studied but much more is known about the composition of the meteorites. If one is to make an estimate of the overall composition of meteoritic material, it is necessary to consider what fraction of a meteorite entering the Earth's atmosphere reaches the surface and can be recovered.

Table 6.   Logarithmic elemental abundances in meteorites.

Element	Abundance	Element	Abundance	Element	Abundance
^{3}Li	3.31 ± 0.04	^{34}Se	3.35 ± 0.03	^{60}Nd	1.47 ± 0.01
^{4}Be	1.42 ± 0.04	^{35}Br	2.63 ± 0.08	^{62}Sm	0.97 ± 0.01
^{5}B	2.88 ± 0.04	^{36}Kr	3.23 ± 0.07	^{63}Eu	0.54 ± 0.01
^{9}F	4.48 ± 0.06	^{37}Rb	2.40 ± 0.03	^{64}Gd	1.07 ± 0.01
^{11}Na	6.31 ± 0.03	^{38}Sr	2.93 ± 0.03	^{65}Tb	0.33 ± 0.01
^{12}Mg	7.58 ± 0.02	^{39}Y	2.22 ± 0.02	^{66}Dy	1.15 ± 0.01
^{13}Al	6.48 ± 0.02	^{40}Zr	2.61 ± 0.03	^{67}Ho	0.50 ± 0.01
^{14}Si	7.55 ± 0.02	^{41}Nb	1.40 ± 0.01	^{68}Er	0.95 ± 0.01
^{15}P	5.57 ± 0.04	^{42}Mo	1.96 ± 0.02	^{69}Tm	0.13 ± 0.01
^{16}S	7.27 ± 0.05	^{44}Ru	1.82 ± 0.02	^{70}Yb	0.95 ± 0.01
^{17}Cl	5.27 ± 0.06	^{45}Rh	1.09 ± 0.03	^{71}Lu	0.12 ± 0.01
^{19}K	5.13 ± 0.03	^{46}Pd	1.70 ± 0.03	^{72}Hf	0.73 ± 0.01
^{20}Ca	6.34 ± 0.03	^{47}Ag	1.24 ± 0.01	^{73}Ta	0.13 ± 0.01
^{21}Sc	3.09 ± 0.04	^{48}Cd	1.76 ± 0.03	^{74}W	0.68 ± 0.02
^{22}Ti	4.93 ± 0.02	^{49}In	0.82 ± 0.03	^{75}Re	0.27 ± 0.04
^{23}V	4.02 ± 0.02	^{50}Sn	2.14 ± 0.04	^{76}Os	1.38 ± 0.03
^{24}Cr	5.68 ± 0.03	^{51}Sb	1.04 ± 0.07	^{77}Ir	1.37 ± 0.03
^{25}Mn	5.53 ± 0.04	^{52}Te	2.24 ± 0.04	^{78}Pt	1.68 ± 0.03
^{26}Fe	7.51 ± 0.01	^{53}I	1.51 ± 0.08	^{79}Au	0.83 ± 0.06
^{27}Co	4.91 ± 0.03	^{54}Xe	2.23 ± 0.08	^{80}Hg	1.09 ± 0.05
^{28}Ni	6.25 ± 0.02	^{55}Cs	1.12 ± 0.02	^{81}Tl	0.82 ± 0.04
^{29}Cu	4.27 ± 0.05	^{56}Ba	2.21 ± 0.03	^{82}Pb	2.05 ± 0.03
^{30}Zn	4.65 ± 0.02	^{57}La	1.20 ± 0.01	^{83}Bi	0.71 ± 0.03
^{31}Ga	3.13 ± 0.03	^{58}Ce	1.61 ± 0.01	^{90}Th	0.08 ± 0.02
^{32}Ge	3.63 ± 0.04	^{59}Pr	0.78 ± 0.01	^{92}U	−0.49 ± 0.04
^{33}As	2.37 ± 0.05				

The logarithm to base 10 of the number of atoms of each element is shown, with uncertainties in the value. The figures are normalised so that the number of silicon atoms is the same as in the solar abudances shown in Table 3. The meteorites lack the volatile gases but provide abundances for many elements not seen in the Sun. Where the two tables both have entries, the majority are very close. (Results are from E Anders, and N Grevesse, *Geochimica and Cosmochimica Acta* **53**, 197, 1989 apart from that for Th which is a later value from E Grevesse and A Noels.)

Meteorites are classified as irons, stones and stony iron, the first type being mainly metallic, the second type mainly rocks and the third type a mixture. All meteorites lose mass from heating in passing through the Earth's atmosphere and the fragment that reaches the Earth in recognisable form may be only a small part of the original meteorite. The irons suffer less than the stones because their heat of vaporisation is higher. Thus the number of irons relative to stones found on the Earth's surface is not representative of the number outside the Earth's atmosphere. Weathering on the Earth is also worse for stones. When allowance is made for heating in the atmosphere and for weathering on Earth, it appears that the most important subgroup of meteorites is one known as carbonaceous chondrites. Chondrites are stony meteorites which contain small spherical bodies known as chondrules embedded in them. It is generally believed that carbonaceous chondrites give the best overall representation of the meteorite material. Their composition is also in good agreement with that deduced for the Earth, apart from volatile gases, and some approximate values are shown in Table 6.

Comparison with solar composition

If one now compares the solar composition given in Table 3 of Chapter 2 with the composition of the Earth and the meteorites, it is found that they are generally similar but with two main exceptions, volatile gases and Li, Be, B. The case of the light volatile gases is easy to understand. If the Earth and the meteorites were formed by aggregation of smaller components, the gases could easily have escaped. There is also little problem with Li, Be, B because of the processes which I have described for other late-type stars in Chapter 6. They are mixed throughout the outer convection zone of the Sun and at the base of this zone the temperature is high enough for some nuclear reactions to occur which destroy Li, Be and B. As a result the Li, Be and B abundances in the Sun should have been reduced relative to the original solar system composition and this agrees with what is found when the solar and terrestrial/meteoritic abundances are compared.

The planets in the solar system can be divided into two main groups, terrestrial (Mercury, Venus, Earth, Mars, (Moon)) and major (Jupiter, Saturn, Uranus, Neptune), with Pluto being anomalous. Many of the satellites of the major planets have similarities to terrestrial planets. The mean densities of the major planets are very much less than the mean densities of the terrestrial planets and this can be understood if they have retained much more of the H and He from the original solar composition. The major planets are, however, believed to have rocky cores comparable in mass with the terrestrial planets. In addition to the planets mentioned, there is of course a horde of minor planets or asteroids mainly situated in orbits between those of Mars and Jupiter.

The composition of objects in the solar system thus appears to be consistent with the view that the system was formed out of a single mass of interstellar gas, although as I have said before the chemical evidence cannot prove that it was. Before turning to a discussion of the solar nebula theory, I have two more comments on the chemical composition.

Cometary composition

Comets have much more eccentric orbits than the other components of the solar system, other than meteor showers which share cometary orbits and are believed to be cometary debris. It is thought that there is a very large reservoir of comets situated at a large distance from the Sun (the Oort cloud, about 2×10^4 AU from the Sun) and moving in much less eccentric orbits. It is believed that from time to time gravitational perturbations due to nearby stars disturb the cometary orbits and cause some to move into the inner solar system and to become new comets as far as terrestrial observers are concerned. Some of these comets are captured by the inner solar system as a result of interactions with the major planets, particularly Jupiter, and they can then become periodic comets like Comet Halley.

The Oort cloud is situated so far from the Sun that some authors have suggested that it is not a permanent member of the solar system but that it has been replenished from time to time since the rest of the solar system was formed. The Galaxy contains giant molecular clouds, which are more dense than the general interstellar medium and which also contain much dust. It is possible that every few hundred million years the Sun and solar system might pass through a cloud. If this happened the Sun might accrete matter from the cloud and this might populate the comet belt. Whether this is possible and whether what would result would be comets is far from clear, but if it did the elemental and isotopic abundances of comets currently entering the inner solar system could be different from those of the rest of the solar system. For this reason there would be great interest in obtaining comet samples for analysis.

Meteoritic anomalies

When the meteorites were first studied, it appeared that all of the meteorites and terrestrial samples had the same isotopic ratios for individual elements as well as the general similarities in overall chemical composition. The general view was summed up by H E Suess in 1965 in an article with the title *Chemical evidence bearing on the origin of the solar system*. He stated:

> Among the very few assumptions which, in the opinion of the writer, can be considered well justified and firmly established, is the notion that the planetary objects, ie planets, their satellites, asteroids, meteorites and other objects in the solar system were formed from a well-mixed primordial nebula of chemically and isotopically uniform composition. At some time between the time of formation of the elements and the beginning of condensation of the less volatile material, this nebula must have been in a state of a homogeneous gas mass at a temperature so high that no solids were present.

This view was changed by the discovery of the Allende meteorite which fell at Pueblito de Allende in Mexico on 8 February 1969. Small inclusions in the Allende meteorite were found to have anomalous isotopic abundances. Because we expect any physical process to treat all isotopes of any element similarly, this suggests that the material of the solar system was not completely well-mixed before the meteorites were formed. One suggestion is that there had been a recent input of newly synthesised material from a supernova explosion into what was about to become the solar system. This idea is reinforced by the discovery of an excess of ^{26}Mg in some of the meteoritic inclusions. The theoretical ideas concerning the formation of the magnesium isotopes require this to have arisen from the decay of ^{26}Al which has a half-life of only 7×10^5 yr. This suggests very strongly that the solar system material received an influx of matter relatively rich in recently formed ^{26}Al just as the solar system was itself forming. Another possible explanation of the inclusions is that small solid particles can survive in the interstellar medium for a very long time and migrate far from their place of origin. In that case, even if the interstellar gas in any region of the Galaxy is kept reasonably well-mixed, the dust may not be so homogeneous and some of it may have what can be regarded as anomalous isotopic abundances. In this case the ^{26}Al, which turns into the observed ^{26}Mg, need not have been produced close to the solar system. Since the discovery of Allende, anomalous abundances have been found in some other meteorites.

The solar nebula theory

The most generally accepted idea relating to the formation of the solar system is that the Sun and planets formed out of a single rotating cloud of interstellar gas. The basic idea is that the central region became the Sun and that the outer part became a flat disk from which the planets formed. In this picture it is natural that the chemical composition of objects in the solar system should be generally similar and that there should also be only a relatively small difference in the ages of the Sun and the planets. The formation of the solar system can then be divided into two stages. The first involves the production of a protosun of something like its present mass and size surrounded by a disk of material in orbit around it. The second is the detailed process of forming planets, satellites, asteroids etc. with their observed properties. Most attention tends to have concentrated on this second phase but the first one is crucially important if only because it provides the initial condition for the second phase.

A basic problem with the observed structure of the solar system is that most of the mass is in the Sun – the present mass of all of the planets etc. is a fraction of one per cent of the solar mass and this would not be increased very much if light gases, which may be presumed to have escaped from the terrestrial planets, are added to make up the solar composition – while most of the angular momentum of the system is the orbital angular momentum of Jupiter about the Sun. If a contracting gas cloud is to form the solar system, somehow this division of mass and angular momentum has to arise. It would require very special initial conditions in the cloud for most of the mass to be essentially non-rotating, while a very small fraction of the mass had significant angular momentum. It therefore seems necessary for the initial disk to evolve with exchange of angular momentum. As I shall explain shortly there are a variety of astronomical objects which contain what are known as accretion disks in which frictional forces cause matter to move inwards on to a central object, the net effect being transport of mass inwards and angular momentum outwards. Such a process could help to produce the solar system but there are remaining problems.

Pre-main-sequence stellar evolution and T Tauri stars

Before discussing the process any further, I should say something about the observational situation. The development of infrared and millimetre astronomy has led to the possibility of observing regions of star formation in the Galaxy and also for finding disks around stars which have already formed. The class of stars known as T Tauri stars, whose activity has already been discussed in Chapter 6, has been recognised for a considerable time as pre-main-sequence stars. They were originally thought to be essentially spherical protostars contracting towards the main sequence, but infrared observations have made it clear that they are surrounded by disks of material, which are thought to be accreting on to the central object. It is not impossible that an important fraction of the mass of the whole star is assembled by accretion from a disk, rather than falling in quasi-spherically.

This does, however, immediately raise a problem both for the T Tauri stars and for the solar system. If a disk of material is in equilibrium about a central object of mass M, the requirement that matter at distance r from M be in circular orbit gives the equation

$$\frac{GM}{r^2} = \frac{v^2}{r}. \tag{8.11}$$

In writing down this equation I have made two assumptions. The first is that the pressure of the disk matter is insignificant. This is necessary for the disk to be thin. The second is that the disk's own mass is negligible. This must in any case be true for the matter which is orbiting close to the surface of M. Accretion disks are observed in cataclysmic variable stars, such as novae, and are also believed to be present in some galactic nuclei and to play a rôle in the quasar phenomenon. In any accretion disk, viscosity is supposed to cause the material to move gradually inwards until it is accreted by the central mass. If equation (8.11) holds during this process, matter arrives with a circular speed comparable with the escape velocity from M ($v_{esc}^2 = 2GM/r$). If much of the mass of a star has arrived by movement through an accretion disk, the star should be rotating very rapidly – i.e. at some significant fraction of its escape or break-up speed. In fact, neither the Sun nor the T Tauri stars are rapidly rotating. This is at odds with the simple conclusion that, although mass and angular momentum can be redistributed in a disk, the process should lead to very rapidly rotating stars.

It is now necessary to introduce another observational fact, which is that outflows of material also appear to be common in regions of star formation. There is clear evidence of outflow of material from T Tauri stars and from other stars in the process of

formation. The prototype objects are called *Herbig–Haro objects* after the two astronomers who first discovered them. The outflows are known as bipolar because a jet of
material is flowing outwards in two opposite directions. What can be observed is the
interaction of the jet with the surrounding interstellar medium. This leads to the
schematic picture which has already been shown in fig. 68. A protostar is surrounded
by a disk of material which is gradually accreting on to the central object. At the same
time there is an outflow of material in directions approximately perpendicular to the
plane of the disk. If the protostar possesses a magnetic field, the loss of mass is
channelled along the field lines. The material ejected by the Sun is ionised and that is
likely to be true in the case of other stars. Such ionised material is strongly tied to the
magnetic field as has been discussed in Chapter 4 and in Appendix 5. I have also
discussed in Chapter 4 how such a magnetically controlled outflow, which persists until
the kinetic energy density of the outflowing matter exceeds the energy density of the
magnetic field, causes magnetic braking of the star. The outflowing material carries away
angular momentum and slows down the stellar rotation. This may be able to offset any
increase in rotation speed due to accretion.

Planetary formation

Suppose that a disk of material finally exists around a main sequence star and that the
process of accretion has become so slow as to be unimportant to the star. What is now
required to form planets? Before I discuss this I should mention some further
observations. The Infrared Astronomy Satellite (IRAS) discovered dusty disks around
several nearby main sequence stars. It is possible that the stars are sufficiently young
that the disks are planetary systems in formation. It is also possible that they are failed
planetary systems, depending on how easy it is to form planetary sized objects from a
circumstellar disk. The best known star to have such a disk is the apparently bright star
Vega. Two stars with particularly massive disks are β Pic, where both optical and
infrared observations have been made, and HR 4796. In the case of β Pic, evidence has
been obtained that the concentration of dust is lower in the regions of the disk nearest
to the star and this may indicate the formation of planets there. Although the disks can
be detected directly it would be much more difficult to detect a planet, because a planet
would appear as a faint point source of reflected starlight close to the star, whereas the
disk appears as a much more extended emitter or absorber of radiation. If a planet is
massive enough, the motion of the star about their centre of mass might however be
observed. More recent observations than IRAS suggest that about a half of all young
stellar objects have disks and that many of those that do not belong to binary systems.
There is such a high fraction of binaries in more mature stars that it is possible that
essentially all young stars are in binary systems or possess disks. A disk has also been
observed around one star in a binary system together with some disruptive influence
due to the second star. At present there is no clear detection of a formed planet around
a normal star. However, as this book is about to go to press, there is a report that a

Jupiter-sized planet has been found with the star 51 Peg. Since then it has been reported that similar planets have been detected around two further stars 70 Vir and 47 UMa. They were discovered because there was a jitter in the star's motion caused by their mutual orbits. There could also be terrestrial-type planets but their effect on the star's orbit would not be observable. Two planetary mass objects have been found orbiting a neutron star. It is not very easy to see how a star can become a neutron star while retaining a planetary system, but if such a system exists, it must be possible. The recent launch of the Infrared Space Observatory (ISO) should lead to very much more detailed information about the properties of star-forming regions.

A recently reported discovery is that of a disk around a solar-type star, or more precisely a main sequence star, which is probably slightly more massive than the Sun. The star is numbered SAO 206462 and it is described as a solar-type star with a dusty organically rich environment. Specifically observations have detected the large organic molecules known as polycyclic aromatic hydrocarbons. These are believed to be an important bridge between gas and dust in the interstellar medium and they have previously been seen in other stars but not in stars similar to the Sun. SAO 206462 has an outer disk of large cold dust particles and an inner disk, which is warmer and which contains smaller dust particles and organic molecules. Obviously, if there are some organic molecules, there may be a smaller number of larger organic aggregates. Although few astronomers (or biologists and chemists) currently agree with the suggestion of F Hoyle and N C Wickramasinghe that life had an interstellar rather than a planetary origin, the evidence will need to be watched carefully.

I now turn to ideas about planetary formation in a solar nebula. The general belief today is that the solar nebula was cold during the early formation phase and that it consisted of a mixture of gas and dust – dust being small solid objects. The solid particles, in the absence of the gas, would move in circular orbits around the central object, the velocity in the orbit being such that the acceleration v^2/r balances the gravitational field due to the star and to the disk itself. The same is not quite true of the gas particles, which would possess a temperature and hence a pressure ($P = nkT$, where n is the particle number density), which would have a gradient in the radial direction and which would partly balance the gravitational force. This means that at any radial position the gas particles have a lower circular velocity than the dust particles. As the solids move through the gas, frictional forces will tend to slow the solids down so that they move with the gas. As a result of this, the gravitational force will exceed the acceleration, v^2/r, and the solids will more inwards. On their new orbits there is an increased probability of collision of solids with solids and of coagulation leading to larger solids.

For a time the growth of solid bodies arises only through collisions and in that case there is always a possibility that collision will produce fragmentation rather than coagulation. However, there comes a stage where the self-gravitation of the individual bodies becomes important. If a stream of small bodies is approaching a large body, in the frame in which the large body is at rest, they will move as in fig. 72. If they collide and

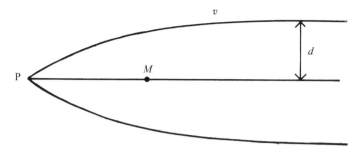

Fig. 72. Accretion by a mass M. Material passing M with speed v is caused to collide at P. The component of velocity perpendicular to the axis is destroyed and the matter may then have less than the escape speed from M.

lose their transverse component of momentum, their velocity will be below escape velocity, if their distance of closest approach to the massive body in the absence of attraction would have satisfied

$$d < 2GM/v^2. \tag{8.12}$$

This implies an increased probability of capture relative to simple collision by a factor d^2/r^2, where r is the radius of the protoplanet, or v_{esc}^4/v^4. This value is really too large because it assumes collisions of particles of equal mass. Even if attention is restricted to particles which are led to collide directly with the protoplanet, the enhancement factor is v_{esc}^2/v^2 for small v and this can lead to a rapid increase in size of the protoplanet. Note that small v is relevant because to a first approximation the protoplanet and the smaller particles are all moving around the star with a velocity given by equation (8.11), so that their difference in velocity is inevitably small. In this way planetary sized objects may have been produced.

So far I have only discussed the manner in which the solids aggregate, but the major component of the solar nebula must be gaseous. One has only to consider the chemical composition of the Sun to realise that this must be the case, if the nebula and the Sun shared the same composition. In general, in the interstellar medium, dust accounts for one or two per cent of the mass. As the protoplanets increase in mass, the possibility arises that they can attract some of the gas gravitationally. This will only occur if the thermal velocity of the gas is less than the escape velocity from the protoplanet. Capture is therefore encouraged by low gas temperatures and by high protoplanetary mass. The density of the solar nebula is expected to be highest in its inner regions and this must encourage the growth of solid bodies. In contrast the gas will have a lower temperature further out and this will encourage the formation of gaseous planets such as the major planets in the solar system.

At any time during the evolution of the nebula there must be two principal factors influencing the temperature of the gas – the conversion of rotational kinetic energy into thermal energy by the viscosity in the disk and irradiation by the protosun. Interstellar gas is not affected very significantly by much of the stellar radiation, which is in the visible and the infrared and which cannot be absorbed by cold hydrogen and helium, which must make up most of the gas. In contrast, all of the radiation interacts with the dust and dust/gas coupling can then heat the gas. Both heating mechanisms will lead to the gas being hotter near to the protosun than further out. We have also seen in Chapter 6 that pre-main-sequence stars tend to be more magnetically active than the present Sun which means that there could have been some additional input of energy into the gas above from the steady solar luminosity.

The general conclusion from discussions of this type is that protoplanets close to the Sun would not be able to accrete and retain a large gaseous envelope. As a result the terrestrial planets (Mercury, Venus, Earth, (Moon), Mars) have a composition which is largely devoid of light gases. This is true in the case of the Earth even though water, made out of such gases, covers most of its surface. The Earth's present atmosphere is believed to have been formed by outgassing from the interior. At a greater distance from the Sun, protoplanets could accrete and retain gas and thus lead to the formation of the giant planets (Jupiter, Saturn, Uranus, Neptune). According to this theory the giant planets must, in effect, contain a terrestrial planet within them. This is different from earlier ideas about the formation of the planets in which they were supposed to have formed by direct condensation of material of solar composition, with the lower mass planets failing to retain most of their light gases. In that case the major planets would not possess a solid core.

The capture theory

The qualitative evidence in favour of planetary formation in a solar nebula is now very strong and it becomes stronger with the discovery of massive disks around young stellar objects. This does not prove that the protosun did automatically have a disk but it seems very likely. For that reason capture theories are now almost completely out of favour. I will, however, say a few words about Woolfson's theory of capture from a protostar. He argues that there are two unsatisfactory features about the solar nebula model which do not occur in his theory and I will discuss these first.

The first is the distribution of angular momentum, with most of the angular momentum of the solar system being possessed by Jupiter. I have argued above that the problem can be solved partially in the solar nebula theory by mass being transferred inwards and angular momentum outwards in an accretion disk but that in addition magnetic braking of the protosun is required to prevent its rotating too rapidly. Woolfson argues that the angular momentum problem does not arise at all in his model and that this is an advantage. His second point is that the rotation axis of the Sun is inclined at an angle of 7 degrees to the normal to the plane of the ecliptic which contains the planets. He says that in a solar nebula theory, it would be expected that the Sun's rotation axis would be perpendicular to the ecliptic, whereas in his model any value of the angle is possible. His objection to the solar nebula theory would be valid, if the Sun had formed from an isolated cloud with a well defined rotation axis and if it did not contain a magnetic field with another axis. It does not, however, seem impossible for an inhomogeneous magnetised cloud to form a Sun with a rotation axis not aligned to that of the disk in which it is embedded.

Woolfson's suggestion is that the Sun was formed in a star cluster and that after it became a main sequence star it had a close passage with a very much larger protostar. As a result it pulled out some material from the protostar and some of this material condensed to form the planets. If this mechanism were possible and were the correct one, it would not contradict anything that I have said earlier in the chapter. Because the Sun is not currently in a cluster this must have dispersed. If this occurred as the result of the explosion of a supernova expelling all of the remaining gas from the cluster and causing the cluster to become unbound, it would be likely to happen when the Sun was not much more than 10^7 yr old, because that is about how long it takes a massive star to become a supernova. This means that the Sun and the planets would have essentially the same age. In addition, all stars in any one cluster tend to have essentially the same chemical composition.

Provided that the Sun could pull sufficient material out of the protostar it would be likely to settle in a plane around the Sun. Although it will not initially be precisely in a plane, the distribution in space must be related to the plane containing the line of centres of the Sun and the protostar and the direction of their relative motion at the time of the encounter. This plane need have no particular relationship with the solar rotation axis. The captured material will collide and dissipate energy and as a result of this dissipation it is likely to move more nearly into one plane and to circularise initially highly eccentric orbits. The material in the disk automatically possesses much more angular momentum than the Sun because the orbital angular momentum per unit mass of the protostar about the Sun is much greater than the spin angular momentum per unit mass of the Sun. The matter which forms the planets is initially orbiting at roughly the radius at which it will settle down. Woolfson and colleagues have performed numerical simulations, which are schematically illustrated in fig. 73, which show the capture process. Once a disk is formed, the process of planetary formation should be similar to that in the solar nebula theory.

The main problem with the capture theory remains its inherent implausibility. Because protostars are so much larger than main sequence stars (Woolfson took a radius of 3×10^{12} m $\approx 4 \times 10^3 r_\odot$) less energy is required to remove material from them. By placing the Sun and a protostar in a cluster the probability of a close passage is increased. Against those advantages, the typical separation of stars in even a dense cluster is of order 30 times the protostar radius and the time allowed for the close encounter to take place is restricted, both by the short lifetime of a star as a protostar and by the brief existence of a bound cluster. If the capture theory were correct planetary systems would be very rare.

It could be argued that this does not matter because we have clear evidence of the existence of only one planetary system. Perhaps we are alone in the Universe. Most astronomers do not take this attitude, at least as far as the existence of planetary

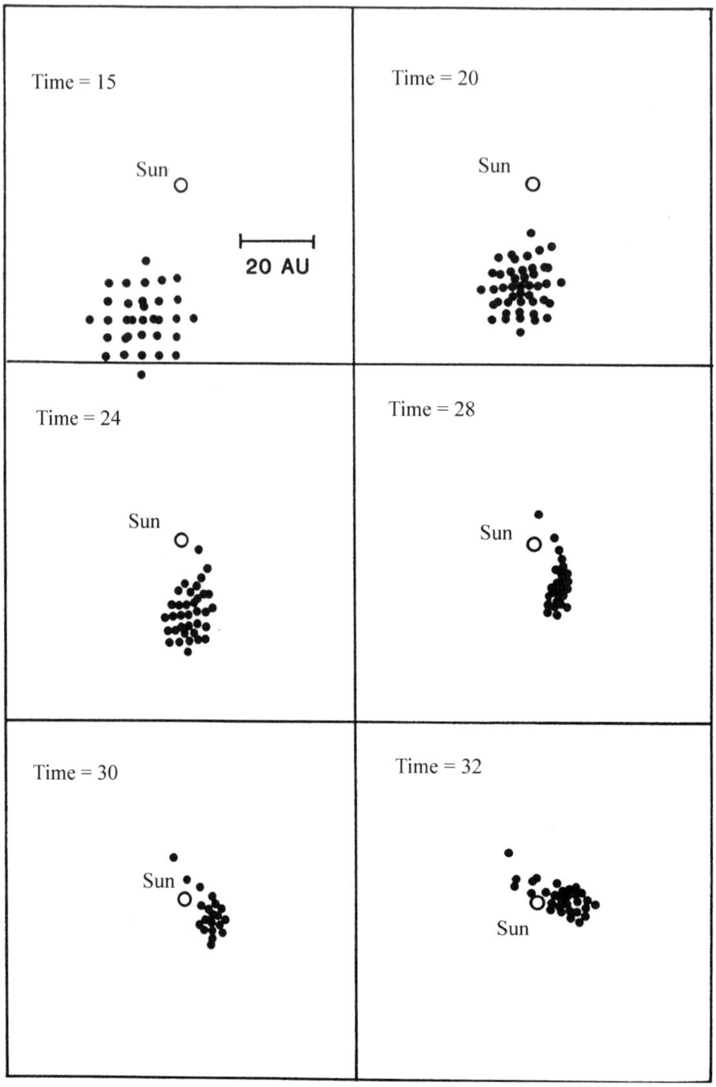

Fig. 73. A sequence of configurations of a tidally disrupted protostar from Dormand, J R & Woolfson, M M, *The Origin of the Solar System: The Capture Theory*. At the end of the calculation most of the protostar is still intact, part has been captured by the Sun and some has become independent of both Sun and protostar. (Courtesy M M Woolfson and Praxis Publishing Ltd.)

systems is concerned. They take the observations of disks about many stars to be evidence that disks are produced as a natural process of star formation. They expect planetary systems to be common in the Galaxy and that many of these planets must possess life or the potential to develop life. One of the more controversial topics is whether advanced life has developed elsewhere but that question is outside the scope of the present book.

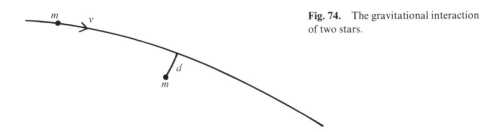

Fig. 74. The gravitational interaction of two stars.

The stability of the solar system

One question in which there is great interest is whether the solar system is stable. In a sense the answer is obvious. Given that the solar system is about 4.6×10^9 yr old and is well-ordered and regular, it must surely be stable. Otherwise it would have broken up long ago. The question can be rephrased. Have the properties of the solar system changed significantly since the end of its formation phase and will they change significantly during the time that the Sun remains a main sequence star? It is clear that, when the Sun becomes a red giant and both its size and its luminosity increase very greatly, the inner solar system at least will be transformed.

The solar system can in principle suffer instability due to external influences or because of its own intrinsic properties. Consider first external influences. I said earlier, in discussing the original interaction theory of the origin of the solar system, that this was inherently implausible because stars in the solar neighbourhood rarely come close enough to one another to influence one another gravitationally. A crude estimate of the frequency of close passages can be obtained as follows. Consider two stars of the same mass m and suppose that one is at rest and that the other moves past it with distance of closest approach, d (fig. 74). The two stars will interact significantly if their mutual gravitational energy at distance d exceeds the kinetic energy. Thus

$$Gm^2/d > \tfrac{1}{2}mv^2. \tag{8.13}$$

If I call d_c the value of d for which there is equality in (8.13), the effective cross-section of a star for collision with another star is

$$\sigma = \pi d_c^2 = 4\pi G^2 m^2/v^4. \tag{8.14}$$

If there are n stars per unit volume, the mean free time between collisions is

$$\tau_c = 1/n\sigma v = v^3/4\pi G^2 m^2 n. \tag{8.15}$$

If I take characteristic values for the solar neighbourhood, $n \simeq 1\ \mathrm{pc}^{-3}$, $m \simeq 1\ M_\odot$, $v \simeq 20\ \mathrm{kms}^{-1}$, the value of τ_c is about 10^{13} yr. It is likely that this crude calculation is not accurate to a factor better than 10, but it is unlikely to give an answer too large by a factor of 1000. It is very improbable that the Sun has had a close interaction with another star since the solar system was formed. This is reassuring as it is difficult to see how a regular solar system could survive such an interaction.

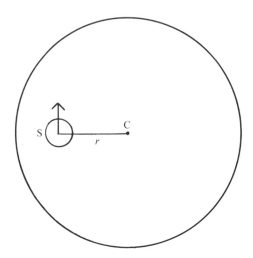

Fig. 75. The passage of the Sun, S, through an interstellar gas cloud with centre C.

Stars are not the only component of the Galaxy and I have mentioned earlier the possibility that the Sun might have passed through a giant molecular cloud. Although such clouds are relatively few in number, they are so much larger than stars that there is a possibility of such a collision. There are perhaps about 10^4 such clouds in the disk of our Galaxy at present which gives an average number density of order $10^{-8}\,\mathrm{pc}^{-3}$. They have a cross-sectional area of order $10^3\,\mathrm{pc}^2$. If I once again take a random velocity of $20\,\mathrm{kms}^{-1}$, which is about $20\,\mathrm{kpc}\,(10^9\,\mathrm{yr})^{-1}$, the value of the collision time is $5\times10^9\,\mathrm{yr}$, which is about the age of the solar system. This is only a crude estimate of the time between collisions with the most massive clouds. There are many more less massive clouds and it is quite likely that the solar system does pass through a cloud every few hundred million years. Would this have a serious effect on its structure?

It is easy to see that in general passage through a gas cloud would not cause many problems. The solar system will be acted on by the gravitational field due to the cloud. Here what is important is the tidal gravitational field which is the difference in gravitational attraction of the cloud on the two sides of the solar system. This needs to be compared with the gravitational attraction of the Sun. If it is assumed that most of the mass of the cloud is within the point where the solar system passes through the cloud (figure 75), that the diameter of the solar system is d, the mass of the cloud is $\lambda M_\odot$ and the solar system is distance r from the cloud's centre, the difference in the gravitational field due to cloud at the two sides of the solar system is

$$\frac{\lambda G M_\odot}{r^2} - \frac{\lambda G M_\odot}{(r+d)^2} \approx \frac{2\lambda G M_\odot d}{r^3}. \tag{8.16}$$

The ratio of this to $GM_\odot/(d/2)^2$ is readily seen to be very small with $\lambda \simeq 10^5$, $r \simeq 10\,\mathrm{pc}$ and d the diameter of the part of the solar system containing the planets. The solar system could clearly survive passage through a gas cloud. A similar argument shows that the tidal gravitational field due to the Galaxy as a whole is unimportant.

Internal stability

A different question is whether the solar system as an isolated object is stable in the sense that the present arrangement of the orbit of the planets and satellites will survive for thousands of millions of years. This is a much more difficult question to answer. In principle it is studied by integration of the orbits of the principal components of the solar system either forward or backward in time. This is not easy because for the inner planets we are concerned with thousands of millions of orbits and it is difficult to ensure that the errors which arise in any numerical calculation do not dominate the results. Direct calculations have at present been performed for about 10^8 yr. It is believed from what has been done so far that there is likely to be and has been significant evolution in the properties of the solar system. Indeed it is suggested that there might have been more small planets in the solar system when it was young and that planetary collisions may have occurred.

Summary of Chapter 8

The problem of the origin of the solar system is related to that of star formation in general. This is a vitally important subject in astronomy. There is beginning to be a wealth of relevant observations but there is not yet a good theoretical understanding of star formation. In particular it is still not known whether or not planetary systems are common, although observations of disks about young stars suggest that they may be.

The ages of the objects in the solar system (meteorites, terrestrial and lunar rocks) can be determined from a study of the radioactive elements which they contain and this suggests that the solar system is about 4.6×10^9 yr old. There is a general similarity in the chemical composition of these objects and of the Sun, when allowance is made for loss of volatile gases by the lighter objects. This is not sufficient to prove that the Sun and the planets were formed at the same time but is consistent with it. There are some isotopic anomalies in meteorites which suggest that the material out of which the solar system formed was not completely well-mixed.

Historically there have been two principal ideas concerning the formation of the solar system. The first is that the whole solar system comes from a rotating cloud of gas, which flattened into a disk and a central Sun as it contracted, and that the planets formed in the disk. The second is that a passing star pulled some material out of the Sun. This material then formed a disk in which planets were formed. It is now recognised that the second model is inherently so improbable as to be untenable and most support is now given to the former, solar nebula, theory. There is a minority opinion in favour of the idea that the Sun pulled the planetary material out of a protostar, which was formed in the same association as the Sun. This is less implausible than the earlier encounter theory but, if it were true, planetary systems should be rare.

The observation of disks around 50 per cent of young stars suggests that the nebula theory is correct. Observations nevertheless indicate that star formation is a complex process. Much of the mass of the central star is accreted from the disk while mass is also ejected in a perpendicular direction in what are known as bipolar outflows. It appears that planets must have originated by the aggregation of small solid objects and that only those planets far enough from the protosun could accrete and retain significant amounts of light gases. This agrees with the distribution of the terrestrial and giant planets.

The solar system has survived about 4.6×10^9 yr and it must therefore be basically stable. It is easy to see that it is unlikely to have collided with another star, which would have been disastrous. In addition, tidal gravitational fields acting during passage through a giant gas cloud or caused by the entire Galaxy, would not have disturbed it. Its own internal stability is less clear and the overall pattern of orbits may have evolved significantly.

9 Concluding remarks

The future of the Sun

In this book I have been principally concerned with the present properties of the Sun and of other active stars, although I have also made some attempt to discuss the Sun's past history. I start this chapter by indicating what astronomers believe will happen to the Sun in the future. This is shown schematically as a path in the Hertzsprung–Russell diagram in fig. 76. The Sun is currently evolving away from the zero age main sequence but gradually increasing in luminosity, but it will continue to be close to the main sequence until it is about twice its present age. There is probably a significant uncertainty in estimates of this time, which is particularly due to doubts about the precise value of the Sun's initial helium content, which most studies put in the range 24 per cent to 28 per cent by mass. The higher is the helium content, the less hydrogen there is to be converted into helium before the Sun leaves the main sequence. Other effects such as lack of a good theory of convection may also be important.

After leaving the main sequence, the Sun should become a *red giant*. All of its post-main-sequence evolution other than its final dying state is believed to take no longer than about 10 per cent of its main sequence lifetime. Detailed predictions become increasingly less reliable, even if the broad outlines are believed to be clear. The ascent of the giant branch should be accompanied by an increasing rate of solar wind mass loss, so that by the time that the Sun reaches the top of the giant branch it may be significantly less massive than it is now. The increasing size and luminosity of the Sun will by that time have transformed if not destroyed the entire solar system. The Earth, in particular, is likely to be engulfed by the Sun. Regardless of any effects that erratic changes in solar luminosity may have on conditions on Earth, the Sun will eventually put an end to life on Earth, if it has not ended of its own accord earlier.

At the top of the giant branch, the central temperature of the Sun should be high enough for nuclear reactions converting helium into carbon, and subsequently oxygen, to start. These reactions are believed to start explosively in stars of masses similar to that of the Sun, for reasons which are explained in *The Stars,* and this leads to a rapid readjustment of a star's internal properties and possibly, but not certainly, to further mass loss. After this *helium flash* the Sun should be on the *horizontal branch*, probably

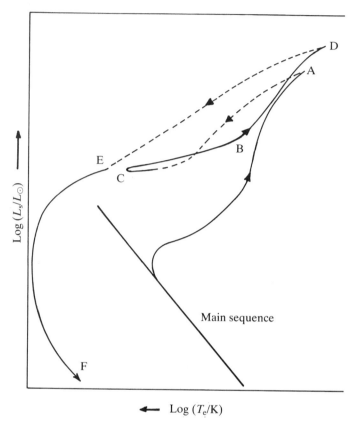

Fig. 76. The evolution of a star of about a solar mass and hence a prediction of the future history of the Sun. After a long period close to the main sequence the star climbs the giant branch to A, where helium ignites explosively. The star joins the horizontal branch probably nearer C than B and climbs the asymptotic giant branch to D, where mass loss produces a planetary nebula. The remaining star becomes a white dwarf which cools from E to F and beyond. The dashed sections represent very rapid evolution.

near the left hand or blue end. The mass of the Sun at this stage will probably be between $0.6M_\odot$ and $0.8M_\odot$. It is then expected to move along the horizontal branch and up the giant branch again along what is called the asymptotic giant branch. By then helium in the central regions has been exhausted and helium burning is occuring in a shell some way out from the centre and hydrogen burning in another shell even further out. Theoretical studies suggest that there will be some instabilities in the helium burning shell and that, as the top of the asymptotic branch is approached, a final instability may remove a significant amount of mass producing what is known as a *planetary nebula*, expanding outwards from a stellar remnant of no more that 0.5 to $0.6M_\odot$.

The stellar remnant should initially be very hot. In fact, some central stars in planetary nebulae are observed to have surface temperatures in excess of 10^5K. Such stars cool very rapidly on an astronomical timescale and they become the dense dying stars known as *white dwarfs* which, as explained in Chapter 2, are the second most common type of star in the HR diagram of nearby stars. The white dwarf would mainly be made of carbon and oxygen. It would eventually decline into invisibility as a black dwarf, but that final cooling could take longer than another 10^{10}yr. There is just a slight possibility that the solar white dwarf could pass through a dense interstellar gas cloud and in so doing could accrete some interstellar hydrogen, which would give it a brief new burst of life, but this is very unlikely.

There is currently considerable uncertainty both in the age of the Universe and in the period for which star formation has been occurring in the disk of the Galaxy. It is possible that the Sun formed as much as 10^{10}yr after the first important period of star formation in the disk of the Galaxy but it is also possible that the time could be as short as 5×10^9yr. If the former is true, stars of a solar mass or a little greater formed in the disk in early galactic history will already be well into their period as cooling white dwarfs. If, in contrast, the latter is true it will be scarcely possible for any solar mass stars to become white dwarfs yet. In the former case, we might hope to be able to observe the whole future history of the Sun by proxy, although even this would be complicated in at least two different ways.

The first is that the chemical composition of stars formed early in galactic history should have been different from that of the Sun and chemical composition may have an important influence on stellar evolution. The second is that the uncertain amount of mass loss at different stages between the main sequence and the final white dwarf state would make it difficult to identify highly evolved stars whose initial mass was that of the Sun. There is good reason to believe that stars whose initial mass is as high as $8M_\odot$ can become white dwarfs of less than $1.4M_\odot$. Because such stars evolve very rapidly, the first white dwarfs would have been produced almost as soon as star formation started. The degree to which observed white dwarfs have cooled towards the final black dwarf state enables an estimate to be made of how long ago galactic star formation started. The result currently obtained suggests that there has not been time for any solar mass star in the solar neighbourhood to become a white dwarf.

The internal structure of the Sun

In the immediate future the main information constraining knowledge of the Sun's internal structure will come from continuing and new solar neutrino experiments and from more detailed study of solar oscillations. Although astronomers continue to have little doubt that the Sun does obtain its radiated energy from nuclear fusion reactions converting hydrogen into helium, it is important for an understanding of stellar structure in general that the reason for the lower than expected neutrino count should be explained. The early evolution of stars which are not much less massive than the Sun is crucial in the determination of the ages of globular star clusters. If the early evolution of the Sun is not fully understood, it is difficult to be certain that the estimated ages of globular clusters are correct. Solar oscillations have the potential to probe the solar interior much more thoroughly than the neutrino experiments. The variation of the solar rotation velocity with depth is one very important piece of information. Constraints on the properties of the convection zone, which are so important for an understanding of magnetic activity, will certainly be improved. In time it is possible that information about any variation of chemical composition with radius will also become available. It would obviously be very valuable to know whether the observed photospheric composition is characteristic of the whole of the Sun, apart from the regions in which nuclear reactions are occurring.

The solar neutrino problem may be resolved by further experiments such as BOREXINO. It is also possible that particle physics experiments may provide information about neutrino masses and neutrino mixing, which could influence interpretation of the solar measurements. Some very ambitious neutrino mixing experiments are currently in the planning stage. The knowledge of the properties of solar oscillations should be greatly increased in the next few years by instruments on the SOHO spacecraft and by a network of ground-based instruments.

From a theoretical point of view, there continues to be the need for a better understanding of convection. Although new theories are being developed which may provide a better description of the process than does the mixing length theory, no one of them has yet received general acceptance and replaced the mixing length theory in most stellar models. Unfortunately the solar convection zone is one in which rotation and magnetic fields interact with convection. It is therefore not clear that a good theoretical description of convection in the absence of rotation and magnetic fields would do any better than the mixing length theory. When the value of the mixing length is adjusted to provide the best fit solar model, this adjustment may implicitly be taking account of rotation and magnetic fields.

The active Sun and other active stars

As far as the Sun itself is concerned, there is no lack of data to be understood. This is not to say that new observations will not lead to new insights. However, there is need for a theoretical understanding of the existing wealth of observations. An obvious example is the heating process in the solar corona. Several mechanisms have been proposed. There is general agreement that the actual process must involve the changing structure of magnetic fields but the precise mechanism, or whether several mechanisms are involved, is still not identified. A large localised energy release produced by a rapidly changing magnetic field structure may be important not only in other active stars and in magnetospheres but also in accretion discs in galactic nuclei. A thorough understanding of both the heating of the solar corona and the energy release in solar flares could therefore have implications for our understanding of other branches of astronomy.

Study of the properties of other active stars is much less advanced than study of the active Sun. Cyclical behaviour in other stars only becomes apparent when observations have been made of many years. Such observations are necessary so that the observational properties of stellar dynamos can be determined as a function of such parameters as the spectral and luminosity classes and the rotational velocity of stars. Only then will the full demands which are placed on theoretical models be known. All the time it must be recognised that solar activity would not be obvious even if the Sun was only as far from us as one of the nearest stars. This implies that there could be a large amount of low level activity on other stars, not only ones similar to the Sun, which we cannot observe.

Techniques for studying other active stars are developing very rapidly. Although direct observations of starspots are not possible, inversion techniques are providing images which make it almost appear that the surface structure is visible. Satellite observations in the X-ray and extreme ultraviolet region of the spectrum by *ROSAT* and EUVE are providing a large amount of new information about stellar activity.

The statement that events observed from a large distance may look simple but that the simplicity is unlikely to survive a close inspection is nowhere more true than in studies of the magnetosphere and the interplanetary medium. Variability is common on almost every spatial and temporal scale. Spacecraft have provided a wealth of observations about the structure of the magnetosphere and about the basic physics of plasmas. Even more data will be provided by the CLUSTER mission. I find it difficult to obtain any clear or simple message from all of these very complex observations. If there is one perhaps it should be a note of warning concerning theoretical studies of plasma phenomena in such environments as supernova remnants and galactic nuclei. Important activity may occur on scales which are smaller than those allowed for in current models.

Another field in which there is very active development is that of the possible existence of other planetary systems. Before the observations made by IRAS there was very little detailed information about the early stages of star formation. Now it is clear that dusty disks are common about young stars and it should not be long before there are many direct images of such disks. These observations give strong support to the general ideas of solar nebula theories for the origin of the solar system and make it seem probable that planetary systems are common. Recent reports of observation of Jupiter-like planets about three stars suggest that it will not be long before there is clear evidence of planetary systems about many normal stars. It should be clear from what I have said that the Sun and the solar system deserves a place right in the centre of modern astronomy. The more we know about it, the more there is to try to understand. It will remain a fruitful research field for a very long time.

Appendix 1 Thermodynamic equilibrium

If a physical system is isolated and left alone for a sufficiently long time, it settles down into what is known as a state of thermodynamic equilibrium. In thermodynamic equilibrium the overall properties of the system do not vary from point to point and do not change with time. Individual particles of the system are in motion and do have changing properties. For example, electrons may be being removed from and attached to atoms. There is, however, a statistically steady state in which any process and its inverse occur equally frequently. Thus, in the example mentioned above, the number of atoms ionised per unit time is equal to the number of recombinations. Because the properties of a system do not vary from point to point when it has reached thermodynamic equilibrium, all parts of it have the same temperature.

If two such isolated systems are brought into contact, heat will flow from one to the other until they reach the same state of thermodynamic equilibrium and hence the same temperature. In thermodynamic equilibrium all of the physical properties of the system (such as pressure, internal energy, specific heat) can be calculated in terms of its density, temperature and chemical composition alone. In nature a true state of thermodynamic equilibrium may be approached closely but never quite reached.

In thermodynamic equilibrium the intensity of radiation is given by the Planck function

$$I_\nu = B_\nu(T) \equiv \frac{2h\nu^3/c^2}{\exp{(h\nu/kT)} - 1}. \tag{A1.1}$$

It should be noted that the Planck function is determined by the temperature alone and does not depend on the density and chemical composition of the material. A particular consequence of thermodynamic equilibrium is Kirchhoff's law

$$j_\nu = \kappa_\nu B_\nu, \tag{A1.2}$$

where j_ν and κ_ν are emission and absorption coefficients for radiation of frequency ν.

Provided corrections due to quantum mechanics and relativity are slight (and this is often true), any species of particle possesses a Maxwellian velocity distribution

$$f(u,\ v,\ w) = n\left[\frac{m}{2\pi kT}\right]^{3/2} \exp\left(-m(u^2 + v^2 + w^2)/2kT\right), \tag{A1.3}$$

where n is the number of the particles in unit volume, m the particle mass, u, v, w the three cartesian components of velocity of the particles and the number of particles in unit volume with velocities between $(u,\ v,\ w)$ and $(u + \delta u,\ v + \delta v,\ w + \delta w)$ is $f\ \delta u\ \delta v\ \delta w$.

In any particular atom or ion, electrons will be arranged in various energy levels. If two such levels have energies E_r and E_s, in thermodynamic equilibrium the numbers n_r, n_s of atoms in the two states will obey the Boltzmann law

$$\frac{n_r}{n_s} = \frac{\exp\left(-E_r/kT\right)}{\exp\left(-E_s/kT\right)}. \tag{A1.4}$$

Finally an atom can exist in several states of ionisation and the number of atoms per unit volume in two successive states of ionisation, n_i,[*] n_{i+1} is related to the electron density n_e by Saha's equation

$$\frac{n_{i+1}n_e}{n_i} = \left[\frac{2\pi mkT}{h^2}\right]^{3/2} \frac{2B_{i+1}}{B_i} \exp\left(-I_i/kT\right), \tag{A1.5}$$

where I_i is the energy required to remove one electron from the atom in the ith state of ionisation and B_i, B_{i+1}, which are called the partition functions for the two states, depend on the electron energy levels in the two ions and the temperature. The chemical composition of the medium enters into Saha's equation because the electrons entering into n_e can be provided by ionisation of all of the elements present.

In the deep interior of a star departures from thermodynamic equilibrium are slight but, as the surface of the star is approached, (A1.1) at least must cease to be true. For one reason, the radiation primarily flows outward and in addition the distribution of radiation with frequency ceases to have the Planck form. If collisions between particles are sufficiently rapid, (A1.2)–(A1.5) may still continue to be valid for some quantity T, which is not a true thermodynamic temperature but which is known as the *kinetic temperature*. If this is true the system is said to be in a state of *local thermodynamic equilibrium*. When population of the states of excitation and ionisation follow (A1.4) and (A1.5), interpretation of properties of spectral lines in terms of element abundances and other stellar properties such as rotational velocity is relatively straightforward. Very near the surfaces of stars it seems certain that the approximations of local thermodynamic equilibrium break down and it is then much more difficult to convert observed line strengths into abundances. This is particularly true of regions where emission lines are formed and that is why stellar emission lines rarely yield reliable abundances.

In the deep interior of the Sun, which is discussed in Chapter 3, conditions are extremely close to thermodynamic equilibrium and all of (A1.1) to (A1.5) can be assumed to be true. In contrast much of solar and stellar activity, discussed in Chapters 5 and 6, which takes place in stellar chromospheres and coronae, occurs in regions where

---

[*] The quantities in (A1.4) should now be written more correctly n_{ir}, n_{is}.

there are significant departures not only from full thermodynamic equilibrium but also local thermodynamic equilibrium. It is however usually safe to assume that there is a well-defined kinetic temperature at which particles have a Maxwellian velocity distribution although there may be a small number of much more energetic particles. Finally, conditions in the interplanetary medium and the solar wind are such that ions and electrons may separately have a distribution which is close to Maxwellian but a different kinetic temperature and there can also be significant departures from the Maxwellian distribution.

Appendix 2 The Zeeman effect

Atomic energy levels and spectral lines

The distribution of electrons in an atom is described in terms of a wave function Ψ which is a function of spatial coordinates and which is such that $|\Psi|^2$ determines the probability that an electron will be in a particular position. The possible electronic energy levels in an atom have wave functions labelled by quantum numbers n, l and m. If the position of an electron relative to the nucleus is described in terms of spherical polar coordinates r, θ, φ, m is associated with the φ part of the wave function, which depends on $\cos m\varphi$ or $\sin m\varphi$. The l quantum number is related to the θ dependence of the wave function and the principal quantum number n to the radial dependence. Although individual quantum mechanical states are labelled by different values of n, l, m, it is possible for several states to possess the same energy.

A simple example is the energy level diagram of hydrogen shown in fig. 77, where the energy levels are labelled by values of n. For hydrogen l and m are integers which satisfy $l < n$ and $|m| \leqslant l$; l is a positive (or zero) integer and m is a positive, negative or zero integer. For given n, there are n^2 quantum mechanical states which have the same energy. Spectral lines arise when an electron moves between two states with different energy by emitting or absorbing a photon.

Zeeman effect

In the presence of a magnetic field (or any other perturbing force which has a preferred direction) the energies of the states with given n and l but different values of m are split. This means that a spectral line can be split into several spectral lines with slightly different frequencies. Consider the Lyα transition of hydrogen. This is a transition between the $n = 1$, $l = 0$ ground state and the $n = 2$, $l = 1$ first excited state. In the presence of a magnetic field the $l = 1$ first excited state splits into three components corresponding to the different values of m, -1, 0 and $+1$. This is shown schematically in

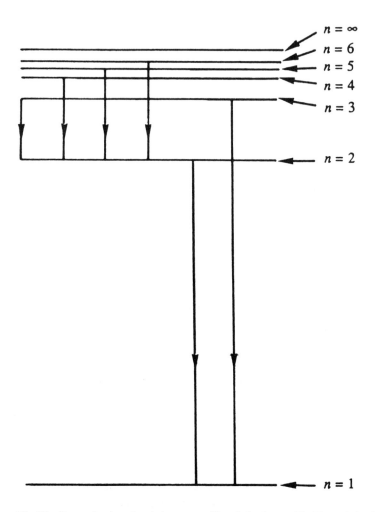

Fig. 77.   Energy levels and emission spectral lines in hydrogen. The Hα emission line is produced by electrons moving from the $n = 3$ level to the $n = 2$ level. Transitions to the $n = 1$ level produce the Lyman series in the ultraviolet.

fig. 78. As a result, in the presence of a magnetic field, the line splits into three components. This is known as the *Zeeman effect*. Note that the transition between $n=2$, $l=0$ and $n=1$, $l=0$ is not involved. This is a *forbidden transition* of a type which will be discussed in Appendix 3. Other spectral lines are much more complicated than this; for Hα , the first line of the Balmer series, there are in principle seven components. For the three component line, the shift in wavelength of the two outer components from the central wavelength is

$$\Delta\lambda/\mathrm{nm} = 4.7 \times 10^{-8} g^*(\lambda/\mathrm{nm})^2(B/\mathrm{T}), \tag{A2.1}$$

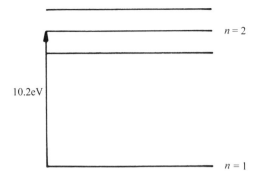

$n = 2$

10.2eV

$n = 1$

Fig. 78. The ground state and first excited state of the hydrogen atom, showing that the first excited state is split into three energy levels in the presence of a magnetic field (not to scale).

where g^*, the modified Landé g factor, depends on the spin and orbital angular momentum of the levels and on the values of m and is usually of order unity.

It is use of the Zeeman effect which enables magnetic fields on the solar surface to be measured. The field has to be very strong as is the field in a sunspot for a splitting to be observed. Even then very high resolution spectroscopy and uncrowded spectral lines are required. Because spectral lines are always broadened for a variety of reasons, if the field is not strong enough the Zeeman splitting is less than the width of the spectral line. In that case what will be observed is a line which is additionally broadened by the Zeeman effect. Then the presence of a magnetic field can only be deduced by deciding that the observed line profile can only be understood in terms of Zeeman broadening. Fortunately the different Zeeman components are polarised in a different manner and this aids in the identification of the effect. The actual splitting that occurs and which line components are observed depends on the orientation of the magnetic field to the line of sight.

For the Sun it is possible to observe a small area of the solar surface over which the field strength and direction may not vary significantly. It must also be recognised that an observed spectral line is not produced at one level in the solar atmosphere but originates from a layer of finite width. If the field is also slowly varying with depth, the observation of Zeeman splitting may be a clear and unambiguous measure of the strength and direction of the field.

For other stars the problem is much more complicated because it is only possible to observe the entire disk of the star. Over the disk it is inevitable that both the magnitude and the direction of the field will vary very greatly. Any observed Zeeman splitting or broadening will be produced by a superposition of lines with different splittings from all over the stellar surface. All that can then be measured is a large scale field and some average value of it. Some stars are observed to have such large scale magnetic fields but the Sun would not appear to have such a field if it were observed from a large distance. As a result the magnetic activity in solar type stars is usually deduced by observations other than a direct magnetic field measurement.

Appendix 3 Atomic energy levels and spectral lines: forbidden transitions

In the simplest quantum mechanical theory of atomic structure only the electrostatic interaction between the nucleus and the electrons is included. Solution of Schrödinger's equation for the wave function Ψ of the system then provides a series of stationary states for which the product $\Psi^*\Psi$, where the star denotes a complex conjugate, is time independent. In this theory the atom can exist indefinitely in any one of these states or energy levels. In the case of the hydrogen atom these states have energies relative to the lowest ($n = 1$) of $13.6\,(1 - 1/n^2)\,\text{eV}$.

In reality all of the states other than the ground state have a finite lifetime. The spontaneous transition from a higher level to a lower level is accompanied by the emission of a photon whose frequency ν is given by $h\nu = E_m - E_n$, where E_n and E_m are the energies of the two levels. The transition arises because an atom has a variable electric dipole moment as well as higher moments such as magnetic dipole and electric quadrupole. A variable dipole moment gives rise to radiation in classical theory and this is usually also the case in quantum theory. The possibility of a transition between states m and n produced by a moment $\mathbf{m}$, which could be an electric dipole or one of the higher multipoles, is determined by the square of the modulus of the integral over all space of $\Psi^*_m \mathbf{m} \Psi_n$, where Ψ_m and Ψ_n are the wave functions for states m and n. In general this probability is much higher for electric dipole transitions than for magnetic dipole transitions and these in turn are more probable than electric quadrupole transitions etc.

When spectral lines were first studied it was recognised that those for a particular element or ion formed series whose frequencies were subsequently shown to obey the relation $h\nu = E_m - E_n$, with n fixed and m variable ($m > n$). In hydrogen we may have Lyman, Balmer, Paschen and Brackett series corresponding to $n = 1, 2, 3, 4$. When such series were studied for some elements and ions, it was found that some of the expected transitions apparently did not occur. They were given the name *forbidden transitions*, although there was no theoretical basis for the use of the word forbidden rather than the more neutral word unobserved. Following the development of quantum mechanics it become clear that the forbidden transitions were those for which the integral of $\Psi^*_m \mathbf{d}_e \Psi_n$ over all space vanished, when $\mathbf{d}_e$ was the electric dipole moment. This means that the probability of radiation due to an electric dipole transition is zero. It was then realised

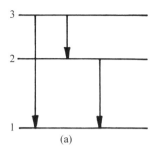
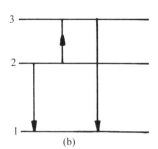

(a)   (b)

Fig. 79. Forbidden transitions. In (a) the transition from 3 to 2 is forbidden and it will not occur because the transition from 3 to 1 is allowed. In (b) the transition from 2 to 1 is forbidden. It will be observed only if collisions are sufficiently infrequent that excitation from 2 to 3 and decay from 3 to 1 or collisional de-excitation from 2 to 1 do not occur first.

that the transition was not forbidden but was extremely unlikely. In general a magnetic dipole transition would be possible. It is, however, very unlikely that it could be observed in a laboratory for two reasons.

The first is that most energy levels have a selection of lower levels to which they can decay. If one of these transitions is an allowed transition (electric dipole radiation), most decays will follow that route (fig 79a). If the forbidden transition is between the first excited state and the ground state, so that no other spontaneous transition is possible (fig. 79b), the second reason is relevant. Collisions between atoms can either put the atom into a higher excited state from which there is an allowed transition to the ground state or cause a collisional transition to the ground state with no emission of radiation. The probability of a forbidden transition is usually so low that collisional excitation and de-excitation occurs first.

In the early years of the 20th century spectral lines, which could not be attributed to any known element, were discovered in the radiation from the solar corona and gaseous nebulae. It was hypothesised that they were produced by elements named *coronium* and *nebulium*. These identifications were encouraged by the fact that helium had been detected in the Sun long before it was found on Earth. It gradually became clear that there was no room in the periodic table of the elements for coronium and nebulium and for a time the origin of the spectral lines was a great puzzle. It was eventually realised that what were being observed were forbidden lines of well known elements; the most prominent nebulium lines arise from low lying levels in twice ionised oxygen. These forbidden transitions can occur in the solar corona and in the nebulae because the particle density is sufficiently low that the competing collisional processes are suppressed. Forbidden lines are usually referred to with their wavelength in square brackets; thus [λ5007] is a forbidden O II (doubly ionised oxygen) line at 5007Å (501 nm). The interpretation of many of the lines in the solar corona was further complicated because they arose from very highly ionised atoms which are only present because of a very high coronal kinetic temperature.

A very important forbidden transition in astronomy, although it is not relevent to the subject of this book, is the 21 cm (0.21 m) radiation of neutral hydrogen. This arises because the ground state is split into two levels in one of which the electron and proton spins are parallel and in the other of which they are antiparallel. The recognition of the existence of this transition gave a great stimulus to the development of radio astronomy and its study provided crucial information about the interstellar medium in our own and other galaxies.

Appendix 4 Theory of solar oscillations

The equations governing the adiabatic oscillations of a spherical star are obtained by keeping only the first order terms in the departure from an equilibrium state in the equations:

Equation of motion $\qquad \rho \dfrac{\mathrm{d}\mathbf{v}}{\mathrm{d}t} = -\mathrm{grad}\, P + \rho\, \mathrm{grad}\, \Phi,^*$ $\qquad\qquad$ (A4.1)

Equation of continuity, $\qquad \dfrac{\mathrm{d}\rho}{\mathrm{d}t} + \rho\, \mathrm{div}\, \mathbf{v} = 0,$ $\qquad\qquad$ (A4.2)

Adiabatic equation, $\qquad \dfrac{1}{P}\dfrac{\mathrm{d}P}{\mathrm{d}t} = \dfrac{\Gamma}{\rho}\dfrac{\mathrm{d}\rho}{\mathrm{d}t},$ $\qquad\qquad$ (A4.3)

Poisson's equation $\qquad \nabla^2 \Phi = -4\pi G \rho.^*$ $\qquad\qquad$ (A4.4)

In these equations Φ is the gravitational potential and $\mathbf{v}$ the fluid velocity. Γ is an effective ratio of specific heats ($\rho\mathrm{d}P/P\mathrm{d}\rho$), which reduces to γ when γ is constant. The time derivative follows the motion of the fluid and is related to the derivative at a fixed point by $\mathrm{d}/\mathrm{d}t = \partial/\partial t + \mathbf{v}.\mathrm{grad}$.

In the equilibrium situation $\rho = \rho_0(r)$, $P = P_0(r)$, $\Phi = \Phi_0(r)$ and $\mathbf{v} = 0$. Small disturbances about this equilibrium are considered in which the perturbed quantities are functions of all the spatial coordinates and the time. Because there is no dependence on spherical polar coordinates θ, φ, or on t in the equilibrium, the dependence of the perturbed quantities on these variables is simple. Thus, if any variable f is written $f = f_0 + f_1, f_1$ has the form

$$f_1 = \mathrm{Re}[\exp(\mathrm{i}\omega_{nl}t)\bar{f}_1(r)Y_l^m(\theta, \varphi)], \qquad\qquad (A4.5)$$

or a similar expression involving derivatives of Y_l^m. In (A4.5), Re means the real part of and $Y_l^m(\theta, \varphi)$ is a *spherical harmonic*

---

* Some authors define the gravitational potential to have the opposite sign. In that case there is a minus sign in equation (A4.1) and a positive sign in (A4.4).

$$Y_l^m(\theta, \varphi) = P_l^m(\theta) \exp im\varphi. \tag{A4.6}$$

Here $P_l^m(\theta)$ is an associated Legendre function. The oscillation frequency does not depend on m for a spherical star. Indeed the dependence on φ does not take the simple form given in (A4.6), if the star is not spherical. However, if the departure from sphericity is small, as is true for the Sun, the real oscillation modes have a behaviour close to that shown in (A4.5) and (A4.6) but with the different m modes having different frequencies. The oscillation frequency ω depends on l and n. The three numbers n, l, m are related to the number of times f_1 vanishes in the radial, θ and φ directions. In addition $|m| \leqslant l$.

The functions $f_1(r)$ now satisfy differential equations which must be solved in order to obtain both expressions for the functions and values for the ωs. In general a fourth order system of equations is obtained and their solution requires four boundary conditions, two at the surface and two at the centre. As in the case of the construction of an equilibrium model of a star, the boundary conditions at the centre of the star are straightforward. There is, however, once again a problem with one of the surface boundary conditions, which arises because stars do not have sharp surfaces. The simplest condition mathematically is to assume that the waves are totally reflected at the solar surface, defined as the level at which density and pressure vanish. In reality some of the wave energy must move outward into the solar atmosphere but, if the reflection boundary condition is applied sufficiently high up in a more realistic atmospheric model, this is probably satisfactory.

I shall not write down the full fourth order system of differential equations governing the perturbed quantities. If the change in the gravitational potential produced by the oscillation is unimportant, the equations reduce to a single second order differential equation. For most perturbations this is a good approximation because some parts of the star are moving inwards and others moving outwards. It is least good for purely radial oscillations. If a perturbation vector $\boldsymbol{\xi}$ is defined by

$$\mathbf{v} = \mathrm{d}\boldsymbol{\xi}/\mathrm{d}t, \tag{A4.7}$$

the single equation can be written in terms of the variable

$$\Psi = c_\mathrm{s}^2 \rho_0^{1/2} \operatorname{div} \boldsymbol{\xi}, \tag{A4.8}$$

where c_s is the velocity of sound in the unperturbed star, $(\Gamma P_0/\rho_0)^{1/2}$. The equation for the radial part of Ψ is

$$\frac{\mathrm{d}^2\psi}{\mathrm{d}r^2} = -\frac{1}{c_\mathrm{s}^2}\left[\omega^2 - \omega_\mathrm{c}^2 - S_l^2\left[1 - \frac{N^2}{\omega^2}\right]\right]\psi. \tag{A4.9}$$

(A4.9) contains three frequencies in addition to ω. These are the acoustic cut-off frequency, ω_c, the Lamb frequency, S_l, and the Brunt–Väissälä frequency, N. They are defined by:

$$\omega_\mathrm{c}^2 = (c_\mathrm{s}^2/4H_\rho^2)(1 - 2\mathrm{d}H_\rho/\mathrm{d}r), \tag{A4.10}$$

where

$$H_\rho = |\rho/(d\rho/dr)| \tag{A4.11}$$

is the density scale height;

$$S_l = c_s[l(l+1)]^{1/2}/r; \tag{A4.12}$$

and

$$N^2 = g\left[\frac{1}{\Gamma P}\frac{dP}{dr} - \frac{1}{\rho}\frac{d\rho}{dr}\right], \tag{A4.13}$$

where

$$g = GM/r^2. \tag{A4.14}$$

It can be seen that S_l is always real but that each of ω_c and N can be imaginary. In particular the condition that N^2 is negative is the condition that convection is occurring as is discussed in Chapter 3.

Equation (A4.9) can be written

$$\frac{d^2\psi}{dr^2} + K_r^2\psi = 0. \tag{A4.15}$$

If K_r^2 were constant, the solutions of equation (A4.15) would depend on its sign. For positive K_r^2, there would be a sinusoidal behaviour with radius. For negative K_r^2 the dependence would be exponential giving an exponentially decaying mode or *evanescent* mode. In reality K_r^2 is a function of r and, for different values of the wave frequency ω, there are regions in the star where the wave propagates and others where it is evanescent. There are two frequency ranges for which K_r^2 is positive. One is that of high frequency waves ($\omega > S_l, \omega_c$) for which pressure fluctuations are most important. These are known as p-modes. The other range is low frequency ($\omega < N$). These are known as gravity or g-modes, because in this case the main restoring force is gravity.

Of course the three frequencies S_l, ω_c and N are themselves functions of position in the Sun. S_l has a simple behaviour; it decreases monotonically from the centre of the Sun to the surface. In a region where convection is occurring N is imaginary. p-modes can propagate in a region whose lower boundary is determined by the Lamb frequency and whose upper boundary, which is close to the surface, is determined by the acoustic cut-off frequency. In contrast the g-modes are trapped beneath the solar convection zone. This means that it is easier to detect p-modes than g-modes at the solar surface, because the g-modes are exponentially attenuated as they travel through the convection zone. The p-modes of lowest order (value of l) penetrate deepest into the Sun. It is they and the g-modes which are capable of providing information about the deep solar interior.

The discussion so far has assumed that the Sun is spherical but, in reality, there are non-spherical effects caused by both rotation and magnetic fields. The discussion of a magnetic field is complicated because there are many free parameters in its possible structure. In addition, the overall influence of the magnetic field is believed to be less than that of rotation and I shall therefore concentrate on the latter. In the absence of any evidence to the contrary, the solar angular velocity, Ω, can be taken to be an arbitrary function of r and θ, where now the axis of the spherical polar coordinate system must be aligned with the rotation axis of the Sun. Because the solar rotation frequency is very small in comparison with the frequencies of the p-modes (rotation period $\simeq$ 28 days, oscillation period $\simeq$ 5 min), a perturbation treatment can be used to discuss the effect of rotation on the oscillations.

For a rotating star the oscillation frequency does depend on m. This means that each frequency ω_{nl} is split into $2l + 1$ frequencies ω_{nlm}. The splitting depends on two things. The first is the way in which the angular velocity varies with position in the star and the second is the variation of the perturbation function with position. Working to first order in Ω the change in frequency produced by rotation can be shown to be

$$\omega_{nlm} - \omega_{nl} = \frac{\int [-m\,\Omega\,\boldsymbol{\xi}.\boldsymbol{\xi}^* + i(\boldsymbol{\Omega} \times \boldsymbol{\xi}).\boldsymbol{\xi}^*]\rho dV}{\int \boldsymbol{\xi}.\boldsymbol{\xi}^* \rho dV}, \qquad (A4.16)$$

where ω_{nlm} and $\boldsymbol{\xi}$ are the non-rotating frequency and perturbation, the asterisk denotes complex conjugate and the integrals are over the whole volume of the Sun. It is easy to check that this expression is real. It is clear from this expression and from what has been said above that modes of different l can give information about the rotation of different parts of the Sun. Thus low l modes, which penetrate close to the solar centre, can provide some information about the whole internal rotation of the Sun whereas high l modes only sample the very outermost layers of the Sun. If the rotational splitting of modes of very different l can be studied, constraints can be obtained on how the rotation velocity varies throughout the solar interior.

The study of the effects of the solar magnetic field on solar oscillations is much more difficult. There are two reasons for this. If the internal field is not much larger than what is suggested by extrapolation of the field in the outer layers, the splitting of frequencies is very small, nHz rather than μHz for rotation. In addition there is no unique direction associated with the magnetic induction vector $\mathbf{B}$ and the effect of $\mathbf{B}$ on the structure of the Sun enters in the same order as its effect on the oscillation frequencies. The most likely progress here is a study of the effect of strong and rapidly varying surface magnetic fields on the oscillations.

In writing this Appendix I have been greatly helped by the article *The Solar Interior* by S Turck-Chièze *et al.* in *Physics Reports*, **230**, 57-235, 1993.

Appendix 5 Motion of a charged particle in uniform electric and magnetic fields

Consider a slowly moving (non-relativistic) particle moving in electric and magnetic fields which are uniform both in space and time. The particle has mass m, charge q and velocity $\mathbf{v}$ and the electromagnetic fields are $\mathbf{E}_0$ and $\mathbf{B}_0$. The equation of motion of the particle is

$$m \frac{d\mathbf{v}}{dt} = q[\mathbf{E}_0 + \mathbf{v} \times \mathbf{B}_0]. \tag{A5.1}$$

I now write

$$\left. \begin{aligned} \mathbf{B}_0 &= B_0\mathbf{b}, \\ \mathbf{E}_0 &= E_{\parallel}\mathbf{b} + \mathbf{E}_{\perp}, \\ \mathbf{v} &= v_{\parallel}\mathbf{b} + \mathbf{v}_{\perp}. \end{aligned} \right\} \tag{A5.2}$$

where $\mathbf{b}$ is a unit vector. Because $\mathbf{v} \times \mathbf{B}_0 \equiv B_0(\mathbf{v}_{\perp} \times \mathbf{b})$ is perpendicular to $\mathbf{b}$, equation (A5.1) immediately splits into two equations

$$m \frac{dv_{\parallel}}{dt} = qE_{\parallel} \tag{A5.3}$$

and

$$m \frac{d\mathbf{v}_{\perp}}{dt} = q[\mathbf{E}_{\perp} + B_0(\mathbf{v}_{\perp} \times \mathbf{b})]. \tag{A5.4}$$

Equation (A5.3) has the immediate solution

$$v_{\parallel} = (qE_{\parallel}/m)t + v_{\parallel 0} \tag{A5.5}$$

where $v_{\parallel 0}$ is the component of velocity in the direction of the magnetic field at time $t = 0$. This ease of particle motions along an electric field parallel to a magnetic field usually leads to the destruction of $E_{\parallel}$; particles of opposite sign move in opposite directions and destroy any charge separation which is responsible for the field. Equation (A5.4) can be solved by writing

$$\mathbf{v}_\perp = \mathbf{v}'_\perp + \mathbf{E}_\perp \times \mathbf{b}/B_0. \tag{A5.6}$$

If (A5.6) is substituted into (A5.5), the left hand side becomes $m\mathrm{d}\mathbf{v}'_\perp/\mathrm{d}t$ while on the right hand side

$$\begin{aligned}
B_0(\mathbf{v}_\perp \times \mathbf{b}) &= B_0\mathbf{v}'_\perp \times \mathbf{b} + (\mathbf{E}_\perp \times \mathbf{b}) \times \mathbf{b} \\
&= B_0\mathbf{v}'_\perp \times \mathbf{b} - \mathbf{E}_\perp,
\end{aligned} \tag{A5.7}$$

by expansion of the triple vector product. Thus equation (A5.4) becomes

$$m\,\frac{\mathrm{d}\mathbf{v}'_\perp}{\mathrm{d}t} = qB_0(\mathbf{v}'_\perp \times \mathbf{b}). \tag{A5.8}$$

This equation now has a simple interpretation. It is motion in which the acceleration is always perpendicular to the velocity and in which the ratio of the acceleration to the velocity is constant. This can only be motion in a circle around the direction of $\mathbf{b}$. The motion has frequency $|q|B_0/m$ so that the period of the circular orbit is $2\pi m/|q|B$. If the magnitude of the velocity is $v_{\perp 0}$, the radius of the orbit, r_g, the gyration radius, is $r_g = mv_{\perp 0}/|q|B$. For an electron the gyration frequency is $1.8 \times 10^{11}\,(B/\text{Tesla})$ Hz. The corresponding gyration radius is $6 \times 10^{-9}\,(v_{\perp 0}/\text{kms}^{-1})/(B/\text{T})$ m. The radius of the orbit and the time taken to describe the orbit are both very small compared to other relevant lengths and times for values of the velocity and magnetic induction observed or inferred in the Sun and the solar system.

The full motion of the particle is thus accelerated motion along the magnetic field, circular motion around the field line and a drift velocity $\mathbf{E}_\perp \times \mathbf{b}/B_0$ perpendicular to both electric and magnetic fields. The sense of the accelerated motion and of the circular motion depends on the sign of the electric charge. The drift velocity is the same for all particles. In the absence of an electric fields, a particle moves with a constant velocity in the direction of the magnetic field and with a velocity of constant magnitude around the field, thus giving a helical path. The simple discussion of a helical orbit will be valid provided that the magnetic induction does not change significantly across the size of an orbit or in the time taken to complete an orbit. This is clearly a very good approximation indeed.

Although this description of charged particles moving in a uniform magnetic field with a constant energy is accurate for a short time, it is not completely accurate. The particle is accelerated in its motion around the field lines and an accelerated charged particle radiates. For non-relativistically moving particles this is known as cyclotron radiation and for relativistic particles synchrotron radiation. Provided, as is usually the case, the fractional energy radiated in any one orbit is minute, the discussion given above is fully adequate. Synchrotron radiation is important in the radiation belts of Jupiter mentioned in Chapter 7.

As a postscript to this discussion, it is convenient to discuss motion in a magnetic field when there is a constant non-magnetic force $\mathbf{F}$ perpendicular to $\mathbf{B}$. With $q\mathbf{E}_0$ replaced by $\mathbf{F}$, the argument is unchanged giving a drift velocity

$$\mathbf{v}_{\text{DF}} = \mathbf{F} \times \mathbf{B}/qB^2. \tag{A5.9}$$

Now v_{DF} is charge dependent and it may also be mass dependent. Thus if $\mathbf{F}$ is gravitational, $\mathbf{F} = m\mathbf{g}$,

$$\mathbf{v}_{DF} = m\mathbf{g} \times \mathbf{B}/qB^2 \tag{A5.10}$$

and the drift velocity depends on the mass/charge ratio of the particle. This principle is used in separating particles with different values of q/m. The ion drift is in this case very much larger than the electron drift. As the particles drift in opposite directions, a current is produced and such drift currents, but produced by a different non-magnetic force, are very important in the Earth's magnetosphere.

Appendix 6 Magnetohydrodynamic waves

In this Appendix I will derive the dispersion relations (4.33) and (4.34) for magneto-hydrodynamic waves in a medium of infinite electrical conductivity and of uniform density and pressure which contains a uniform magnetic field. The results will also be essentially correct for non-uniform media provided that the properties of the medium only change in distances large compared to the wavelength $(2\pi/k)$ and in times large compared to the period $(2\pi/\omega)$. The frequency ω is supposed to be sufficiently low that the displacement current is much smaller than the conduction current. It is implicitly being assumed that $2\pi/\omega$ is also very small compared to the resistive decay time $\tau_D = \mu_0\sigma L^2$. All of these assumptions are valid in most circumstances.

The basic equations are then

$$\rho\frac{\mathrm{d}\mathbf{v}}{\mathrm{d}t} = -\operatorname{grad}P + \mathbf{j}\times\mathbf{B}, \tag{A6.1}$$

$$\frac{\mathrm{d}\rho}{\mathrm{d}t} + \rho\operatorname{div}\mathbf{v} = 0, \tag{A6.2}$$

$$\frac{1}{P}\frac{\mathrm{d}P}{\mathrm{d}t} = \frac{\gamma}{\rho}\frac{\mathrm{d}\rho}{\mathrm{d}t}, \tag{A6.3}$$

$$\operatorname{curl}\mathbf{B} = \mu_0\mathbf{j}, \tag{A6.4}$$

$$\frac{\partial\mathbf{B}}{\partial t} = \operatorname{curl}(\mathbf{v}\times\mathbf{B}). \tag{A6.5}$$

I can immediately insert (A6.4) into (A6.1) to obtain

$$\rho\frac{\mathrm{d}\mathbf{v}}{\mathrm{d}t} = -\operatorname{grad}P + (\operatorname{curl}\mathbf{B}\times\mathbf{B})/\mu_0. \tag{A6.6}$$

The relevant equations are now (A6.2), (A6.3), (A6.5) and (A6.6).

In equilibrium $\rho = \rho_0$, $P = P_0$, $\mathbf{B} = B_0\mathbf{b}$, where $\mathbf{b}$ is a unit vector, and $\mathbf{v}$ is zero. I consider perturbations about the equilibrium in which

$$\rho = \rho_0 + \rho_1 = \rho_0 + \rho_{10}\exp\mathrm{i}(\mathbf{k}\cdot\mathbf{r} - \omega t), \tag{A6.7}$$

and similarly for P, $\mathbf{B}$ and $\mathbf{v}$, where ρ_{10} etc. are constants. Eventually it is the real part of this expression for ρ which is relevant. Because I am interested in small amplitude waves,

I insert expression (A6.7) and the similar expressions for P, $\mathbf{B}$ and $\mathbf{v}$ into the basic equations and retain only terms which are linear in the small deviations from equilibrium. Because there is no $\mathbf{v}$ in the equilibrium, and no spatial variation of either ρ_0 or P_0, I can immediately replace d/dt by $\partial/\partial t$. Because of the exponential dependence of the perturbed quantities,

$$\operatorname{grad} P_1 = iP_1\mathbf{k}, \quad \operatorname{div} \mathbf{v}_1 = i\mathbf{k}.\mathbf{v}_1, \quad \operatorname{curl} \mathbf{B}_1 = i\mathbf{k} \times \mathbf{B}_1. \tag{A6.8}$$

The first order equations can then readily be seen to take the form

$$-i\omega\rho_1 + i\rho_0\mathbf{k}.\mathbf{v}_1 = 0, \tag{A6.9}$$

$$i\omega P_1 = (\gamma P_0/\rho_0)i\omega\rho_1, \tag{A6.10}$$

$$-i\omega\mathbf{B}_1 = i\mathbf{k} \times (\mathbf{v}_1 \times B_0\mathbf{b}), \tag{A6.11}$$

$$-i\omega\rho_0\mathbf{v}_1 = -iP_1\mathbf{k} + [(i\mathbf{k} \times \mathbf{B}_1) \times B_0\mathbf{b}]/\mu_0. \tag{A6.12}$$

The first three equations give ρ_1, P_1 and $\mathbf{B}_1$ in terms of $\mathbf{v}_1$

$$\rho_1 = (\rho_0/\omega)\mathbf{k}.\mathbf{v}_1, \tag{A6.13}$$

$$P_1 = c_{\mathrm{s}}^2\rho_1 = (c_{\mathrm{s}}^2\rho_0/\omega)\mathbf{k}.\mathbf{v}_1, \tag{A6.14}$$

$$\mathbf{B}_1 = -(B_0/\omega)\mathbf{k} \times (\mathbf{v}_1 \times \mathbf{b}), \tag{A6.15}$$

where, as previously defined, $c_{\mathrm{s}}^2 = \gamma P_0/\rho_0$. If these equations are inserted into (A6.12) there results

$$\mathbf{v}_1 = (c_{\mathrm{s}}^2/\omega^2)\mathbf{k}.\mathbf{v}_1\mathbf{k} + (B_0^2/\rho_0\omega^2\mu_0)[\mathbf{k} \times (\mathbf{k} \times (\mathbf{v}_1 \times \mathbf{b}))] \times \mathbf{b}$$
$$= (c_{\mathrm{s}}^2/\omega^2)\mathbf{k}.\mathbf{v}_1\mathbf{k} + (c_{\mathrm{H}}^2/\omega^2)[\mathbf{k} \times (\mathbf{k} \times (\mathbf{v}_1 \times \mathbf{b}))] \times \mathbf{b}, \tag{A6.16}$$

where as before $c_{\mathrm{H}}^2 = B_0^2/\rho_0\mu_0$. The very complicated multiple vector product in the final term can then be expanded to give the result

$$\mathbf{v}_1 = (c_{\mathrm{s}}^2/\omega^2)(\mathbf{k}.\mathbf{v}_1)\mathbf{k} + (c_{\mathrm{H}}^2/\omega^2)[(\mathbf{k}.\mathbf{b})^2\mathbf{v}_1 - (\mathbf{k}.\mathbf{b})(\mathbf{b}.\mathbf{v}_1)\mathbf{k}$$
$$-(\mathbf{k}.\mathbf{v}_1)(\mathbf{k}.\mathbf{b})\mathbf{b} + (\mathbf{k}.\mathbf{v}_1)\mathbf{k}]. \tag{A6.17}$$

This is now three equations coupling the three components of $\mathbf{v}_1$, which implies that the resulting dispersion relation is a cubic equation in ω^2.

A first simple result can be obtained by taking the scalar product of (A6.17) with $\mathbf{k} \times \mathbf{b}$. Because triple scalar products with two vectors the same vanish identically, the equation

$$\mathbf{v}_1.\mathbf{k} \times \mathbf{b} = (c_{\mathrm{H}}^2/\omega^2)(\mathbf{k}.\mathbf{b})^2\mathbf{v}_1.\mathbf{k} \times \mathbf{b}, \tag{A6.18}$$

results. From this either $\mathbf{v}_1 . \mathbf{k} \times \mathbf{b} = 0$ or

$$\omega^2 = c_{\mathrm{H}}^2(\mathbf{k}.\mathbf{b})^2. \tag{A6.19}$$

Two further equations can be obtained by taking the scalar products of (A6.17) with $\mathbf{k}$ and $\mathbf{b}$ respectively. Thus

$$\mathbf{k}.\mathbf{v}_1 = k^2(c_{\mathrm{s}}^2/\omega^2)\mathbf{k}.\mathbf{v}_1 + k^2(c_{\mathrm{H}}^2/\omega^2)[(\mathbf{k}.\mathbf{v}_1) - (\mathbf{k}.\mathbf{b})(\mathbf{b}.\mathbf{v}_1)] \tag{A6.20}$$

and

$$\mathbf{b}.\mathbf{v}_1 = (c_s^2/\omega^2)(\mathbf{k}.\mathbf{v}_1)(\mathbf{k}.\mathbf{b}). \tag{A6.21}$$

Insertion of equation (A6.21) into equation (A6.20) gives either $\mathbf{k}.\mathbf{v}_1 = \mathbf{b}.\mathbf{v}_1 = 0$ or

$$\omega^4 - k^2(c_s^2 + c_H^2)\omega^2 + k^2 c_s^2 c_H^2(\mathbf{k}.\mathbf{b})^2 = 0,$$

the dispersion relation for magnetosonic waves.

Appendix 7 Dynamo maintenance of magnetic fields

In Chapter 4 I considered the interaction of a magnetic induction $\mathbf{B}$ with a given velocity field $\mathbf{v}_0 + \mathbf{u}$ where $\mathbf{v}_0$ was slowly varying in space and time and $\mathbf{u}$ was a fluctuating velocity with a zero mean value. If $\mathbf{B}$ was similarly written $\mathbf{B}_0 + \mathbf{b}$, $\mathbf{B}_0$ and $\mathbf{b}$ satisfied the equations

$$\frac{\partial \mathbf{B}_0}{\partial t} = \mathrm{curl}\,(\mathbf{v}_0 \times \mathbf{B}_0) + \mathrm{curl}\,\langle \mathbf{u} \times \mathbf{b} \rangle - \mathrm{curl}\,(\eta\,\mathrm{curl}\,\mathbf{B}_0), \tag{A7.1}$$

$$\frac{\partial \mathbf{B}}{\partial t} = \mathrm{curl}\,(\mathbf{v}_0 \times \mathbf{B}_0) + \mathrm{curl}\,(\mathbf{u} \times \mathbf{B}_0) + \mathrm{curl}\,(\mathbf{u} \times \mathbf{b} - \langle \mathbf{u} \times \mathbf{b} \rangle)$$
$$- \mathrm{curl}\,(\eta\,\mathrm{curl}\,\mathbf{b}), \tag{A7.2}$$

where I have now made it explicit that the resistivity may vary in both space and time. In (A7.1) and (A7.2), $\langle\ \rangle$ denotes a mean value.

If $\mathbf{b}$ is assumed to be initially zero, the term $\mathrm{curl}\,(\mathbf{u} \times \mathbf{B}_0)$ in (A7.2) is the generator of the subsequent $\mathbf{b}$ field. With the velocities $\mathbf{v}_0$ and $\mathbf{u}$ specified, equation (A7.2) is linear in $\mathbf{b}$ and $\mathbf{B}_0$, which means that there should be a linear relationship between $\mathbf{b}$ and $\mathbf{B}_0$ and its derivatives or equivalently between $\langle \mathbf{u} \times \mathbf{b} \rangle$ and $\mathbf{B}_0$. Since the characteristic length scale of variation of $\mathbf{B}_0$ is supposed large compared to that of $\mathbf{u}$ and $\mathbf{b}$, it can be hoped that we can write

$$\langle \mathbf{u} \times \mathbf{b} \rangle_i = \alpha_{ij} B_{0j} + \beta_{ijk}\,\frac{\partial B_{0i}}{\partial x_k} + \gamma_{ijkl}\,\frac{\partial^2 B_{0i}}{\partial x_k \partial x_l} \cdots \tag{A7.3}$$

and that only the first few terms in this series will be important. In (A7.3), if a suffix is repeated, summation over that suffix is implied. Thus

$$\alpha_{ij} B_{0j} = \alpha_{i1} B_{01} + \alpha_{i2} B_{02} + \alpha_{i3} B_{03} \tag{A7.4}$$

where the suffixes 1,2,3 refer to x, y and z components respectively and $x_1 = x$, $x_2 = y$, $x_3 = z$. In principle the expansion (A7.4) could include time derivatives of $\mathbf{B}_0$ but these can be eliminated using equation (A7.1).

Consider the first term in (A7.3). All of the coefficients in (A7.3) are determined from the fluctuating velocity $\mathbf{u}$. If the $\mathbf{u}$ field is statistically isotropic and homogeneous, then there cannot be any preferred direction for the α term so that inevitably

$$\alpha_{ij} = \alpha\delta_{ij}, \tag{A7.5}$$

where δ_{ij} is the Kronecker delta which is unity if $i = j$ and zero otherwise. Note that α is what is known as a pseudo-scalar because in any change from a right-handed to a left-handed coordinate system, $\langle \mathbf{u} \times \mathbf{b} \rangle_i$ does not change sign whereas B_{0i} does. This then leads to the final term in my expression (4.46) for $\langle \mathbf{u} \times \mathbf{b} \rangle$ Note that $\langle \mathbf{u} \times \mathbf{b} \rangle$ appears in the equations as an effective electric field parallel to the mean magnetic field and this leads to current parallel to the magnetic field.

The appearance of an electromotive force $\alpha \mathbf{B}_0$ was given the name α *effect*. It is the key term in dynamo theory as was mentioned in Chapter 4. The essence of any dynamo maintenance of a magnetic field is that it must be possible both to produce a toroidal field from a poloidal field and to produce a poloidal field from a toroidal field. The first step is easy. This occurs if differential rotation acts on any poloidal field. The effect with a current parallel to the poloidal field produces a toroidal field. It was this that was not possible in the cases considered by Cowling in his anti-dynamo theorem. It is not difficult to show that α can only be non-zero if the $\mathbf{u}$ field lacks reflexional symmetry; this means that the statistical properties of the field are changed if $\mathbf{x}$ is changed to $-\mathbf{x}$.

The β term can be discussed in a similar manner. If the $\mathbf{u}$ field is homogeneous and isotropic β_{ijk} must also be isotropic. It takes the form

$$\beta_{ijk} = \beta\epsilon_{ijk}, \tag{A7.6}$$

where $\epsilon_{ijk} = 1$ if $(ijk) = (123)$ or an even permutation of it and -1 for an odd permutation. The term in β then becomes $-\beta$ curl $\mathbf{B}_0$ as in equation (4.46). It is expected that as $\mathbf{B}_0$ is supposed to be slowly varying in space and time, the terms in γ_{ijkl} in (A7.3) will be less important than the α and β terms.

If the statistical properties of $\mathbf{u}$ and the form of $\mathbf{B}_0$ are specified, it is possible to solve equation (A7.2) to obtain expressions for α and β. The derivation is far beyond the level of the present book. If the fluctuating fluid motions are ones of turbulent convection expressions of the forms given in equations (4.47) and (4.48) can be obtained. There is however no completely general expression which applies in all circumstances and dynamo theory remains a subject which is still incompletely understood.

Appendix 8 Plasma properties

Plasma oscillations

Suppose a system of electrons and ions is initially at rest. I will consider the case of pure hydrogen but the discussion can easily be generalised. Suppose that the equilibrium is disturbed so that a non-zero charge distribution is produced. Because protons are much more massive than electrons, it is much easier to move the electrons and in my discussion I assume that electrons move relative to protons at rest. I suppose that a sinusoidal distribution of electric charge is set up in the direction of the x coordinate. The equations which I need to consider are:

1. *Equation of motion of electrons:*

$$m_e \frac{d\mathbf{v}_e}{dt} = -e\mathbf{E}. \tag{A8.1}$$

This is simply Newton's law.

2. *Equation of continuity for electrons:*

$$\frac{\partial n_e}{\partial t} = -\text{div}\,(n_e \mathbf{v}_e). \tag{A8.2}$$

This equation is a consequence of a vector formula known as the *divergence theorem*. From fig. 80, the number of electrons leaving the volume V per unit time is $\int n_e \mathbf{v}_e.d\mathbf{S}$, where the integral is over the surface surrounding V. The divergence theorem states that

$$\int n_e \mathbf{v}_e.d\mathbf{S} = \int \text{div}\,(n_e \mathbf{v}_e)dV, \tag{A8.3}$$

where the second integral is through the volume V. But this must also be equal to the rate of decrease in the number of electrons in volume V or to

$$-\int \frac{\partial n_e}{\partial t}\,dV. \tag{A8.4}$$

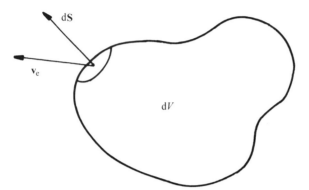

Fig. 80.   The number of electrons inside a volume is changed as a result of electrons crossing the surface.

If the second term in (A8.3) is equated to (A8.4) and it is recognised that the result must be true for an arbitrary volume V, the equation of continuity results.

3.   *Poisson's equation:*

$$\epsilon_0 \operatorname{div} \mathbf{E} = -\rho_e. \tag{A8.5}$$

There is a slight subtlety in the equations of motion and continuity. $\partial/\partial t$ is the rate of change with time at a fixed point in space but d/dt the rate of change with time following the motion of an electron. It can be shown that $d/dt \equiv \partial/\partial t + \mathbf{v}_e \cdot \operatorname{grad} = \partial/\partial t + v_{ex}\partial/\partial x$, if there is motion only in the x direction. In my later discussion the two derivatives are the same to first order in small quantities because $\mathbf{v}_e \cdot \operatorname{grad} \mathbf{v}_e$ contains the square of v_{ex} which is small.

Suppose that in my equilibrium

$$n_e = n_p = N \quad \text{and} \quad \mathbf{v}_e = \mathbf{E} = \mathbf{0}, \tag{A8.6}$$

where n_p is the number of protons per unit volume and N is a constant. Suppose further that the equilibrium is perturbed so that

$$n_e = N + n \exp(ikx - i\omega t), \tag{A8.7}$$

$$v_{ex} = v \exp(ikx - i\omega t), \tag{A8.8}$$

$$E_x = E \exp(ikx - i\omega t), \tag{A8.9}$$

n_p is unchanged, v_{ey}, v_{ez}, E_y, E_z are zero and the physical quantities are the real parts of the above expressions. I can put these into equations (A8.1), (A8.2) and, because I am considering small disturbances, I keep only those terms which are linear in small quantities. If I do that, the equations become (omitting the dependence $\exp(ikx - i\omega t)$)

$$-i\omega m_e v = -eE, \tag{A8.10}$$

$$-i\omega n = -ikvN, \tag{A8.11}$$

$$ik\epsilon_0 E = -ne. \tag{A8.12}$$

These are now three algebraic equations, which only have non-zero solutions if ω satisfies a simple relation

$$\omega^2 = Ne^2/\epsilon_0 m_e \equiv \omega_{pe}^2, \tag{A8.13}$$

where ω_{pe} is the *electron plasma frequency*. This shows that electrons in an ionised gas can oscillate with a natural oscillation frequency ω_{pe}. Note that the oscillation frequency does not depend on k. This means that coherent oscillations can exist through any region in which N is effectively constant. The oscillations are known as *plasma oscillations*. If the motion of the protons is included, plasma oscillations occur at a frequency ω_p given by

$$\omega_p^2 = Ne^2/\epsilon_0 m_p + Ne^2/\epsilon_0 m_e. \tag{A8.14}$$

Because $m_p \gg m_e$, $\omega_p \approx \omega_{pe}$.

Electromagnetic waves in a plasma

In discussing electromagnetic waves, I make the same assumption that only the electrons move and that the ions are at rest. Electromagnetic waves are transverse waves unlike plasma oscillations which are longitudinal. This means that, if the wave propagates in the x direction, the electric and magnetic fields and the electron velocity are in the y, z plane. In consequence div $\mathbf{E}$ and the charge density are both zero. The equations that are to be solved are equation (A8.1) and two of Maxwell's equations, which were discussed in Chapter 4,

$$\text{curl } \mathbf{B} = \mu_0 \mathbf{j} + \mu_0 \epsilon_0 \frac{\partial \mathbf{E}}{\partial t}, \tag{A8.15}$$

$$\text{curl } \mathbf{E} = -\frac{\partial \mathbf{B}}{\partial t}. \tag{A8.16}$$

I can now write

$$\mathbf{j} = -Ne\mathbf{v}_e, \tag{A8.17}$$

and insert this into (A8.15). If I then differentiate (A8.15) with respect to t and use equations (A8.16) and (A8.1), with d/dt replaced by $\partial/\partial t$, there results

$$\nabla^2 \mathbf{E} + \frac{\omega_{pe}^2}{c^2} \mathbf{E} = \frac{1}{c^2} \frac{\partial^2 \mathbf{E}}{\partial t^2}. \tag{A8.18}$$

To obtain (A8.18), I have also used the vector identity curl curl $\equiv$ grad div $- \nabla^2$ and $(\mu_0\epsilon_0)^{-1} = c^2$. If $\mathbf{E}$ is assumed proportional to $\exp{(ikx - i\omega t)}$, the result

$$\omega^2 = \omega_{pe}^2 + k^2 c^2, \tag{A8.19}$$

is obtained. This shows that, for given ω and large enough ω_{pe}, k becomes imaginary and

E is attenuated as a function of x rather than propagating. For smaller ω_{pe} the propagation speed is changed.

It is easy to understand why electromagnetic waves cannot propagate if ω_{pe} is large enough. As an electromagnetic wave tries to enter a plasma, the plasma electrons move in the field of the wave so setting up a field which opposes the wave field. The electrons can react to the wave field at the frequency ω_{pe} and, if this frequency is greater than the wave frequency, they are capable of cancelling out the wave field. If, in contrast, $\omega_{pe} < \omega$, the electrons cannot keep up with the wave fluctuations and the wave propagates with somewhat modified properties. This linear treatment assumes that there are enough electrons present to produce a large enough electric field. If the wave field is very strong, this may not be the case. The wave then picks up the electrons and accelerates them to high energies. This is one possible mechanism for the production of relativistic particles.

There are two velocities associated with the electromagnetic wave, the phase velocity ω/k which is greater than c and the group velocity $d\omega/dk$, which is the velocity with which energy and information travel and this is always less than c in accordance with the principle of relativity.

Debye length

The discussion of the particle collision frequency given in Chapter 7 is over simplified for two reasons. The first arises because the electrostatic coulomb force is long range. In a uniform medium the fall off with $1/r^2$ of the strength of the force is balanced by the increase with r^2 of the number of particles at distance r. At first sight the contribution of interactions with distant particles or *distant collisions* should completely dominate the close collisions, except for the occasional interaction with a particle which approaches much closer than the mean interparticle separation. This is essentially true in the case of self-gravitating systems of stars because the gravitational force is always attractive but another factor is important in systems which carry no net electric charge. This is the effect known as *Debye screening*.

Consider a spherical region centred on a particle P (fig. 81). Suppose P carries a negative electric charge. It will attract all of the positive electric charges which are close to it and it will repel all of the negative charges. As a result of this cloud of positive charges which surrounds it, its electrostatic influence will be screened out beyond a certain distance, which is known as the *Debye length*. The consequence of this is that a proper treatment of collisions need only consider particles which are up to about a Debye length apart.

A proper derivation of the expression for the Debye length is very complicated but it is not difficult to give a plausible explanation of the form that it takes. Thus the Debye length, λ_D, is given by

$$\lambda_D^2 = \epsilon_0 k_B T / n_e e^2 = (k_B T / m_e) / (n_e e^2 / \epsilon_0 m_e) = v_{th}^2 / \omega_{pe}^2, \tag{A8.20}$$

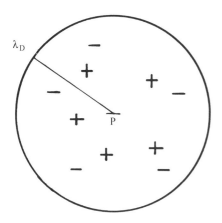

Fig. 81. A schematic illustration of Debye screening. A negative electric charge at P attracts positive charges and repels negative charges. As a result its influence is screened out at a distance λ_D.

where in this appendix the Boltzmann constant is k_B to avoid confusion with wave number k. The final expression in (A8.20) contains a typical thermal velocity of an electron and it indicates that λ_D is the typical distance travelled by an electron in the time of a plasma oscillation. The plasma frequency provides a measure of the ability of electric charges to move to screen out an electrostatic field. Within the distance λ_D, the random thermal motions of the particles can counteract this tendency but at greater distances the random particle speeds do not enable them to travel far enough to prevent the Debye screening. If the full quantitative discussion is given, it is found that the effective electrostatic potential of a charge e is not $e/4\pi\epsilon_0 r$ but is

$$\varphi = \frac{e}{4\pi\epsilon_0 r} \exp\left(-2^{1/2} r/\lambda_D\right). \tag{A8.21}$$

As discussed in Chapter 7, this Debye screening is very important in the design of experiments to study the properties of the space plasmas close to the Earth.

The effect of thermal motions on plasma properties

In my discussion of plasma oscillations I have assumed that initially the electrons and ions are at rest. This is the approximation of a *cold plasma*. In fact there will always be thermal motions, as I have recognised in my description of the Debye length. If thermal motions are included, the plasma frequency is no longer independent of the wavelength of the wave and to a first approximation equation (A8.13) is replaced by

$$\omega^2 = \omega_{pe}^2 + 3k_B T k^2 / m_e = \omega_{pe}^2(1 + 3k^2\lambda_D^2), \tag{A8.22}$$

which looks very similar to equation (A8.19) for the propagation of electromagnetic waves in a plasma, but with the velocity of light replaced by a thermal velocity. The standard treatment of plasma oscillations should therefore be valid if the wavelength of the oscillations is much greater than the Debye length. Even this is not the full story. It is also found that the waves suffer a damping known as *Landau damping*, so that the oscillations have a characteristic lifetime. The fine details of what correction term should be included in equation (A8.22) and of the damping depend on whether or not the particles possess a Maxwellian velocity distribution.

Appendix 9 Motion of a charged particle in a spatially varying magnetic field

The magnetic mirror effect

If a particle of mass m and charge q moves a time-independent magnetic field, the equation of motion (for non-relativistic motion, $v \ll c$)

$$m \frac{d\mathbf{v}}{dt} = q\mathbf{v} \times \mathbf{B} \tag{A9.1}$$

can easily be integrated to show that the particle's kinetic energy is a constant of the motion. Thus take the scalar product of equation (A9.1) with $\mathbf{v}$. This gives

$$m\mathbf{v}.\frac{d\mathbf{v}}{dt} = \frac{d}{dt}\left[\frac{1}{2}mv^2\right] = 0, \tag{A9.2}$$

so that $\frac{1}{2}mv^2$ is constant. This can be written

$$v_\perp^2 + v_\parallel^2 = \text{const}, \tag{A9.3}$$

where $v_\perp$ and $v_\parallel$ are components of the velocity perpendicular to and parallel to $\mathbf{B}$. The above result can be generalised to show that the total energy, rest mass plus kinetic, of a relativistically moving particle is constant. Although the lack of variation with time of $\mathbf{B}$ has not been used explicitly in the above proof, it is implicit because, if $\mathbf{B}$ varies with time, Maxwell's equations show that there must be an electric field present which should be included in (A9.1).

Consider now the explicit case of motion in a converging magnetic field (fig. 82). The magnetic field is to a first approximation in the z direction but the radial component becomes important as the region of high field is approached. The magnetic field must satisfy

$$\text{div } \mathbf{B} = 0 = \frac{1}{r}\frac{\partial}{\partial r}(rB_r) + \frac{\partial B_z}{\partial z}. \tag{A9.4}$$

Assume that an individual particle orbit is so small that $\partial B_z/\partial z$ is essentially constant across it. Also, as $B_z \approx B$ the magnitude of $\mathbf{B}$, I can write it $\partial B/\partial z$. Then I can integrate equation (A9.4) to get

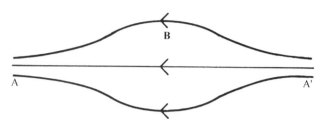

Fig. 82.   A converging magnetic field producing a magnetic mirror.

$$B_{\mathrm{r}} \approx -\frac{r}{2}\frac{\partial B}{\partial z}. \tag{A9.5}$$

On the orbit of a particle moving around the central field line $r \approx mv_\perp/qB$. Thus the parallel component of equation (A9.1) can be written

$$m\frac{\mathrm{d}v_\parallel}{\mathrm{d}t} = qv_\perp B_r = qv_\perp\left[-\frac{mv_\perp}{2qB}\frac{\partial B}{\partial z}\right] = -\frac{mv_\perp^2}{2B}\frac{\partial B}{\partial z}. \tag{A9.6}$$

The quantity $mv_\perp^2/2B \equiv \mu$ is called the *magnetic moment* of the charged particle; it is simple to check that it has the dimensions of a magnetic moment. The equation can therefore be written

$$m\frac{\mathrm{d}v_\parallel}{\mathrm{d}t} = -\mu\frac{\partial B}{\partial z}. \tag{A9.7}$$

Multiply (A9.7) by $v_\parallel$. The left hand side becomes $\mathrm{d}/\mathrm{d}t\left[\frac{1}{2}mv_\parallel^2\right]$. The right hand side is

$$\mu\frac{\partial B}{\partial z}v_\parallel = -\mu\frac{\partial B}{\partial z}\frac{\mathrm{d}z}{\mathrm{d}t} = -\mu\frac{\mathrm{d}B}{\mathrm{d}t}, \tag{A9.8}$$

where $\mathrm{d}B/\mathrm{d}t$ is just the rate of change of B following the motion of the particle. I thus have

$$\frac{\mathrm{d}}{\mathrm{d}t}\left[\frac{1}{2}mv_\parallel^2\right] = -\mu\frac{\mathrm{d}B}{\mathrm{d}t}, \tag{A9.9}$$

but from equation (A9.3)

$$\frac{\mathrm{d}}{\mathrm{d}t}\left[\frac{1}{2}mv_\parallel^2\right] = -\frac{\mathrm{d}}{\mathrm{d}t}\left[\frac{1}{2}mv_\perp^2\right] \equiv -\frac{\mathrm{d}}{\mathrm{d}t}(\mu B). \tag{A9.10}$$

Comparing these last two results, I have

$$\mathrm{d}\mu/\mathrm{d}t = 0, \tag{A9.11}$$

so that the magnetic moment does not change as the particle moves. This is not a completely rigorous proof, because of the approximations which I made, but a more rigorous treatment shows that the departure of μ from constancy is very small indeed.

   If μ is constant, $v_\perp^2$ must increase as B increases. But, since $v_\perp^2 + v_\parallel^2$ is constant, eventually $v_\parallel$ may vanish and the particle motion will reverse. This is the magnetic mirror effect. It is important in the Earth's radiation belts.

Gradient B drifts

Suppose that I have a magnetic field in the z direction whose value varies with y but not with x (fig. 83). Thus

$$\mathbf{B} = B(y)\hat{\mathbf{z}}, \tag{A9.12}$$

where $\hat{\mathbf{z}}$ is a unit vector in the z direction. Since B varies in the y direction, as the particle's motion takes it into the region where the field is weaker, the instantaneous value of its gyration radius, $v_\perp m/|q|B$, gets larger. As a result of this, the orbit does not close but a drift of the centre of the orbit is set up as shown in fig. 83. For the simple case illustrated this *gradient B drift* can be shown to be

$$\mathbf{v}_{\nabla B} = \pm \frac{1}{2} v_\perp r_{\mathrm{g}} (\partial B_z / \partial y)\hat{\mathbf{x}}, \tag{A9.13}$$

where $\hat{\mathbf{x}}$ is a unit vector in the x direction and the plus sign applies to electrons and the minus sign to ions. r_{g} is the gyration radius. The general expression is

$$\mathbf{v}_{\nabla B} = \mp \frac{v_\perp r_{\mathrm{g}}}{2} \frac{\mathbf{B} \times \nabla \mathbf{B}}{B^2}, \tag{A9.14}$$

where now the minus sign applies to electrons. Because charges of opposite sign drift in different directions, a current is set up and this in turn produces a magnetic field which tends to oppose the original field. In the Earth's magnetosphere, gradient B drift

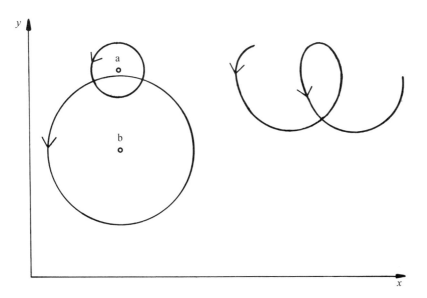

Fig. 83. Motion of a charge in a spatially varying magnetic field. The field into the paper is stronger at a than at b giving a smaller gyration radius there. As a particle tries to adjust to the local value of the gyration radius, its motion is as shown at the right.

produces a *ring current*, which circles the Earth and one of the consequences of a magnetic storm is a change in the ring current.

Curvature drift

There is yet another drift to be taken into account in studies of the magnetosphere. This arises because the dipolar magnetic field lines are curved and it is called *curvature drift*. We can expect the particles to move in tight spirals around the magnetic field lines (fig. 84). Because a particle is moving on average on a curved trajectory (motion of its guiding centre), it feels an outward centrifugal force, $\mathbf{F_c}$. At the point A in the figure, the local radius of curvature of the field line is R_c so that I have

$$\mathbf{F_c} = (mv_\parallel^2/R_c)\hat{\mathbf{r}}, \tag{A9.15}$$

where $\hat{\mathbf{r}}$ is a unit vector in the outward radial direction. In Appendix 5 I discussed drift velocities produced by non-magnetic forces. I can use the formula given there to obtain

$$\mathbf{v}_{\mathrm{curv}} = \mathbf{F_c} \times \mathbf{B}/qB^2 = mv_\parallel^2\hat{\mathbf{r}} \times \mathbf{B}/R_c qB^2. \tag{A9.16}$$

The drift is perpendicular to $\hat{\mathbf{r}}$ and $\mathbf{B}$ and it depends in magnitude among other things on the kinetic energy of the particle parallel to $\mathbf{B}$, $\frac{1}{2}mv_\parallel^2$, and in direction on the sign of the particle's charge. Curvature drift also contributes to the ring current.

In the Earth's radiation belts, the drift velocities are much slower than the velocities of particles between their mirror points but their effect is to cause particles to occupy a spherical shell rather than to move up and down a single field line.

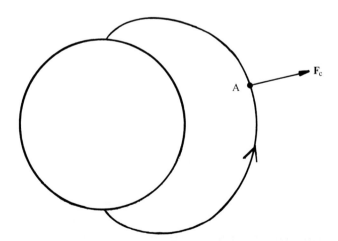

Fig. 84. Curvature drift. A particle moving in a tight spiral along a curved magnetic field line experiences a centrifugal force $\mathbf{F_c}$ shown which produces a curvature drift.

Suggestions for further reading

There are many books on the Sun. Three good books which are entirely non-mathematical are:

> R Kippenhahn, *Discovering the Secrets of the Sun*, John Wiley & Sons.

> R W Noyes, *The Sun, Our Star*, Harvard University Press.

> K J H Phillips, *Guide to the Sun*, Cambridge University Press.

All of these books discuss solar energy as a power source on Earth in addition to more purely astronomical matters. Phillips has a very brief introduction to activity in other stars. Another non-mathematical book which is brief and very fully illustrated is:

> R Giovanelli, *Secrets of the Sun*, Cambridge University Press.

More advanced treatments of the Sun include

> C J Durrant, *The Atmosphere of the Sun*, Cambridge University Press.

> M Stix, *The Sun*, Springer-Verlag.

> H Zirin, *Astrophysics of the Sun*, Cambridge University Press.

Of these, Stix's book is largely a theoretical treatment, while the other two concentrate more on the observations. A theoretical text on the properties of solar magnetic fields is:

> E R Priest, *Solar Magnetohydrodynamics*, D Reidel.

The theory of stellar structure is discussed in:

> R J Tayler, *The Stars: Their Structure and Evolution*, Cambridge University Press.

A very full treatment of the solar neutrino problem can be found in:

> J N Bahcall, *Neutrino Astrophysics*, Cambridge University Press.

The properties of active stars are not at present well covered in textbooks, so that the main source of information is review articles and conference proceedings, such as those published in *Annual Review of Astronomy and Astrophysics*. One text is:

> P R Wilson, *Solar and Stellar Activity Cycles*, Cambridge University Press.

I know of no simple up-to-date textbook on solar–terrestrial relations but standard textbooks include:

J K Hargreaves, *The Upper Atmosphere and Solar Terrestrial Relations*, Van Nostrand.

G K Parks, *Physics of Space Plasmas: An Introduction*, Addison Wesley.

A very recent multi-author text is:

M Kivelson and C T Russell (eds), *Introduction to Space Physics*, Cambridge University Press.

The solar nebula theory of the origin of the solar system is also not covered in texts at an appropriate level but a vast amount of information is contained in:

E H Levy and I L Lunine (eds), *Protostars and Planets 3*, Arizona University Press.

This is a successor to:

D C Black and M S Matthews (eds), *Protostars and Planets 2*, Arizona University Press.

T Gehrels (ed), *Protostars and Planets: Studies of Star Formation and of the Origin of the Solar System*, Arizona University Press.

For an alternative view on the origin of the solar system see:

J R Dormand and M M Woolfson, *The Origin of the Solar System: The Capture Theory*, Ellis Horwood.

The age of the solar system is discussed in:

R J Tayler, *The Origin of the Chemical Elements*, Wykeham Publications.

Index